From the Atacama to Makalu

A Journey to Extreme Environments

on Earth and Beyond

R. T. Arrieta

Coquí Press

Panama City, Florida

From the Atacama to Makalu
A Journey to Extreme Environments on Earth and Beyond

by R. T. Arrieta

Published by:

Coquí Press
Post Office Box 32083
Panama City, FL 32407-8083 U.S.A.

Publisher's Cataloging in Publication
(*Prepared by Quality Books Inc.*)

Arrieta, R. T.
From the Atacama to Makalu : a journey to extreme environments on Earth and beyond / R.T. Arrieta.
p. cm.
Includes bibliographical references and index.
Preassigned LCCN: 96-71973
ISBN 0-9655967-5-3

1. Ecophysiology. 2. Biogeography. 3. Extreme environments. 4. Exobiology. I. Title.

QH541.15.E26A77 1997 574.5
QBI97-40171

CONTENTS

Preface

The silverfish that scurry around in your garage (especially if it is full of cardboard boxes as mine is) are marvels of adaptation, although most people hardly think of this when chasing them around the corners in an insecticidal frenzy. Their peculiar talent is their ability to get by, and even thrive, without any water either to drink or in the starch they eat. How is this possible? What other marvels of evolution exist in our cupboards? My garage is a limiting environment, but there are many other more exotic limiting environments.

The limits to all Earth-based life can be couched in terms of availability of water, heat, a carbon source such as CO_2, an electron acceptor such as oxygen, and an energy source such as light. These factors interact with each other in complex ways, and the interactions are different for different environments. To display and highlight these interactions, I have chosen a unique set of extreme environments. This set is hardly comprehensive, but each environment in the set exaggerates one (or sometimes two) of the factors to such an extent that it becomes the dominant force driving adaptation in that ecosystem. This book is meant to explore those extreme environments from an ecophysiological perspective.

No book has yet been written on the limits to life as they apply to all creatures. There are erudite tomes on individual aspects of the limits to life and advanced college texts that deal with these aspects as they apply to individual groups of organisms like plants or invertebrates. But none of those books is written at the introductory college level. This book is meant to fill this gap; it is aimed at college students, high school students taking advanced science classes, and any educated layperson interested in the subject of extreme environments on Earth and elsewhere.

My fascination with harsh environments dates from the days of the Viking missions to Mars in the late seventies. Since then, I have searched for examples of the limits to life on Earth, hoping to find unifying principles behind these limits.

I have written this book with several objectives in mind. One of them is to draw attention to Earth's extreme environments and the creatures that call such places home. Many times, extreme environments are the last to generate conservation efforts, for in our society there is a deep-rooted bias against wastelands. I think they are worth preserving, not least because of what they teach us about the endurance limits of Earth-based life.

Another objective is to give a sense of biogeography to the reader, something that is sadly lacking in most biological curricula in the United States. In this respect, the book is a compendium of biogeographical information about places few of us will ever visit. Since the book is as much about biogeography as it is about physiological ecology, I have, where appropriate, quoted from the original naturalists-explorers who first ventured into these extreme environments. Their inquisitiveness and amazement still echo in their words after decades, or even centuries, and effectively convey what makes science such a human endeavor.

A third objective is for this book to serve as a layperson's introduction to physiological ecology. It explains the different limits organisms encounter in harsh environments. Some of the mechanisms that those organisms have evolved to deal with those limits are simple but elegant, while others are intricate and clever. To really understand why those environments are harsh, one must study the physiology of the organisms involved; but physiology only yields half the picture as far as survival is concerned. No organism is an island, and even in simple ecosystems, interrelationships between organisms are as important as the environmental factors in determining which organisms thrive and which do not. For instance, even the toughest plant may need a pollinator, and limitations on pollinators then translate into limitations on the plant.

This book is also meant to stimulate the imagination and expose the possibilities for life on other planets; therefore, the last chapter examines the unique challenges that would face life on Mars and how hypothetical organisms would meet those challenges. Mars and the Earth have often been called sister planets. From their beginning, they diverged: one to a successful career in the life business and the other to a desolate and austere death (maybe). On Mars there are several possible places where indigenous life may have hidden or at least have left traces of its previous existence. There are also several possible adaptations found in terrestrial life that may approximate the adaptations necessary for an organism to survive somewhere on Mars. All these possibilities take your breath away, but their development traditionally has been left to science fiction writers. These possibilities have other uses besides serving as plot devices, and I believe they can be effective educational tools. The progress of science is tied to the imagination of scientists and the realization of possibilities. The last chapter of this book outlines these possibilities, and it is meant to stimulate the imagination of budding scientists and to give science teachers food for thought as they prepare projects linked to the forthcoming exploration of Mars. I therefore hope this book will find a place in suggested reading lists for courses in the natural sciences.

I fully expect a flurry of activity in exobiology over the next few years, as data from NASA's Mars Global Surveyor orbiter and Mars Pathfinder lander trickle back to Earth. These data will not only supply us with much needed information on the Martian surface but also spur new research on our own planet. It is my hope that this book will help the researchers of the future find answers in unobvious places.

Because one of my aims is to teach a little biogeography along the way, I have included three appendices. They are meant to help readers find the material about a particular organism or location and also to give them a sense of the evolutionary relationships between the organisms. Most of the uncommon terms used in the book are explained in the glossary. The list of references is organized by chapter, and it is meant to serve as an introduction to the topics covered in a particular chapter.

Why write this book now? I feel now is the time to share this fascinating chapter of life on Earth, before we embark again on a quest for life on Mars. We've recently celebrated the twentieth anniversary of the Viking landings (July through September 1996). When I started writing the book over a year ago, I thought Mars would be in the public eye as the anniversary approached, but I had no inkling of the work being done at the Johnson Space Center on the Allen Hills meteorite. As this book goes to the printer, the Mars Global Surveyor orbiter and the Mars Pathfinder lander are on their way to Mars and other launches will occur in 1998 and 1999. I hope those events, and this book, will spur interest in physiological ecology on our planet among teachers and students alike.

Before about 3.8 billion years ago Mars had flowing water in some form. Why did it stop flowing and when? The scars left by these flows are dramatic evidence, but there are few equivalent erosional features on Earth from which to draw conclusions. Was liquid water present on Mars long enough to allow life to evolve? Maybe. The questions are overwhelming, and at the moment the answers are few. One thing is certain, however, if cryptic life exists or sleeps on Mars, our goal should be to find it and learn from it.

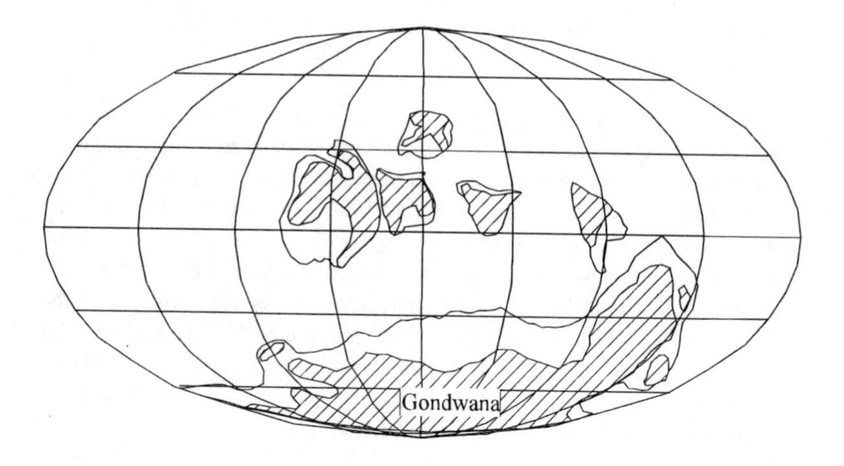

The world as it may have appeared in the Upper Silurian. During this time, land organisms became common enough to later appear as fossils. The hatched outline represents land; the clear outline represents shallow water. (Estimated from the Mollweide projections of Scotese, et. al. 1979. *Jour. Geology*, 87, no. 3, 217-277.)

1 The Extremes of the Past

Life as we now know it began on Earth about 3.6 billion years ago. What went on before then is the subject of intense speculation and debate. What "life" existed back then? It was not species and groups of related species (genera), at least not the way we define these concepts today. If a proto-organism made copies of itself, was this act reproduction? How much error can there be in a copy before it ceases to be a copy? There must have been a time, in the giddy days of simple molecular systems, when an individual could be its own order, class, and phylum. The first individuals that exerted quality control over their copying processes and, with this enhanced copying ability, generated reasonable facsimiles of themselves, could take over the world. Once some proto-organisms discovered the virtues of packaging and compartmentalization, they gained an advantage over their naked neighbors. The original packaging was lipid micelles for some and hydrocarbon micelles for others. These micelles were armored vessels, and the helmsman was by now DNA. Cellular life had evolved. The hapless naked protos must have been still plentiful, but they had no chance against these newly armored, moving fortresses. Some fortresses were benign, harvesting only the rich, disorganized soup around them. Some, however, developed extracellular digestive enzymes—contact mines, as it were. These contact mines were cast into the water. When an unsuspecting proto ran into one, it meant disintegration, oblivion. And its innocuous remnants were assimilated. The sea was made a desert, and only the armored fortresses were left. The arms race that ensued has been going on ever since.

Initially Earth's atmosphere lacked oxygen, as can be seen by the lack of reddish rocks (stained by oxidized iron) over 2 billion years old. This fact has two profound implications for the history of life on Earth. First, the lack of oxygen meant that the surface of this planet was bathed in short ultraviolet (UV) radiation. These energetic UV rays can dissociate water as well as other chemical compounds, producing extremely reactive and therefore biologically damaging molecules called free radicals. The sterilizing effects of this UV radiation must have produced a truly barren land surface, and even in the oceans, the deadly UV probably penetrated 10 meters down. Second, the early organisms would have been extremely sensitive to oxygen. We breathe an oxygen mixture, and so does every other creature on Earth whether it flies, swims, or slithers on the surface of the planet, and plants make oxygen with abandon. What happened to change the status quo so drastically? Blue-green algae happened. These organisms, more than any other, irrevocably changed the course of life on this planet. They were the first to come up with a new process, oxygenic photosynthesis; in effect, they invented molecular fire. Just as fire changed the course of human development, the buildup of oxygen in the atmosphere changed the course of evolution on this planet.

Before the invention of oxygenic photosynthesis, a few types of bacteria had harnessed light energy to maintain their internal electron-rich state (relative to the outside world), but the process they used wasted much of the energy available from light and used raw materials that were not always available. The blue-greens not only came up with a way to extract more of the energy available in light photons, but they also used a ubiquitous raw material to maintain their internal electron-rich state. This raw material was water. Remove two electrons from water and you get a very reactive oxygen atom. The blue-greens invented a process by which they took two of those very reactive oxygen atoms, joined them to form the dimeric molecule that is gaseous oxygen, and excreted this gas as the waste product of their new process. This unregulated toxic waste must have caused havoc in the bacterial communities the blue-greens belonged to. The speed with which the bacterial communities adapted to this threat is anyone's guess, but adapt they did. The presence of oxygen allowed new biochemical processes to evolve, and some claim that these processes were crucial to the eventual evolution of eukaryotic cells and, later, metazoans.

After the invention of oxygenic photosynthesis by blue-green algae, as the concentration of oxygen in the atmosphere increased, the harmful effects of short UV radiation began to decrease. On the one hand, oxygen is a very reactive molecule, and it must have been poison to the creatures of the early Earth, as it still is to many modern bacteria. An early organism unlucky enough to drift into an area of high oxygen concentration stood little chance of survival. On the other hand, when UV photons of the right energy strike an oxygen molecule, the oxygen-oxygen bond is torn asunder, leaving extremely reactive monatomic oxygen. This monatomic oxygen can react with another oxygen molecule to become ozone, and ozone is a good absorber of short UV radiation. The screening of deadly UV radiation opened the door to the invasion of the land by plants and animals alike. Up to that time in Earth's history, the surface of the land had been the most extreme environment of all.

By the time the ozone shield was thick enough to allow wholesale invasion of the land (in the Silurian period), many organisms must have been using the boundary between the land and the sea as a convenient refuge from predation. Such an adaptation can be seen today in the horseshoe crab (*Limulus polyphemus*). This ancient chelicerate excavates a shallow depression on the shore at the level of the highest spring tides and deposits its eggs in it. Horseshoe crabs may have been performing this ritual for the past 400 million years—something to ponder next time you observe

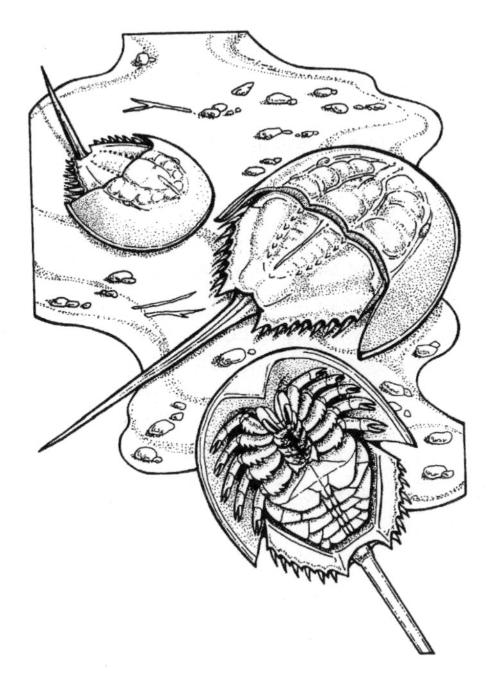

The horseshoe crab (*Limulus polyphemus*)

their telltale telson marks on the sand. Before the advent of terrestrial life, the shore beyond high tide was a haven where organisms could deposit their eggs, and as long as they were covered by some sand, they were safe from the multitude of predators in the sea. A clutch of eggs is a concentrated source of nutrition in any epoch. Once such concentrated sources of nutrition were available beyond the surf, it was only a matter of time before other organisms ventured forth to collect this bounty. Physical agents such as wind, surf, and storms must also have delivered substantial amounts of organic matter onto the shores and beyond, making the land suitable for those small creatures that could spend a whole generation on one piece of wrack. Scientists who have studied the present-day versions of these aeolian (wind-supplied) ecosystems have commented time and again on the preponderance of primitive forms in these ecosystems. This is an often-used but little-deserved appellative, since, as this book will show, some of those primitive organisms have been very successful. For example, springtails dominate the aeolian ecosystems of today, and it is easy to imagine that they have done so since the Silurian period, when they first evolved (in fact, their fossils are some of the oldest terrestrial arthropod fossils).

Many authorities do not classify springtails as insects, but since they are undeniably related groups, I will label them proto-insects in this book. The oldest insect-like fossil ever found, predating true insect fossils by 50 million years, is the springtail *Rhyniella praecursor*, discovered in the Old Red Sandstone of Britain, which dates from the middle Devonian period, 350 million years ago. This fossil is almost indistinguishable from modern springtails, and it shows how unjust it is to label as "primitive" a successful evolutionary design lasting over a third of a billion years.

It is important to note that labèling a species as "primitive" does not attribute to it a lack of adaptation. When a species is referred to as "primitive" because of similarity to a fossil, the comparison is necessarily morphological, and not all adaptation is accomplished through morphological means, or indeed even reflected in morphological attributes. There are several other aspects of adaptation that are just as important as form and shape, but are not as well preserved (if at all) in the fossil record. Adaptation can be achieved through behavioral, biochemical, and physiological means. A small amount of behavioral information is available from the fossil record if the animals were buried rapidly after death, and if the site remained fairly undisturbed until excavated. With modern techniques, more and more biochemical information is retrievable from relatively recent fossils (less than 35 million years old). As far as physiological information, little can be gained aside from those clues given by morphology. Even though morphological similarity is of necessity the yardstick by which evolutionary relationships to extinct species are measured, biochemical evolution and physiological evolution assure that a currently successful species will be as advanced as its environment requires it to be.

2 Aridity and its Effect on Life

Introduction

With no water, no life can exist; but very little water, or water at infrequent intervals, goes a long way. Terrestrial plants and animals are an evolutionary tribute to the importance of water. The exploitation of super-dry environments could not proceed in earnest until vascular plants came onto the stage. The advances seen in vascular plants with respect to water conservation are many, from fertilization mechanisms that no longer require an external wet surface to water-storage organs. Animals faced similar challenges and developed their own solutions. In the first half of this chapter we will explore the challenges encountered in arid environments as well as the biological advances developed to cope with these challenges. The second half of the chapter explores those regions of the planet where these adaptations are found.

When we think of deserts we normally think of dry and dusty stretches of smoky seared shrubs and, maybe, succulents. But there are other environments that are desert-like with respect to the availability of water. Sometimes simply looking at the plants that grow in these environments is enough to convince a person that such an environment is a desert. The canopy of a drought-deciduous forest, where some epiphytic cacti grow, is such a place. These cacti, along with other plants, perhaps found competition at the level of the soil too intense and took to the trees. Some of the other epiphytes have given up roots altogether and absorb moisture through their leaf surfaces. In the Americas, these cryptodeserts among the branches can be found from the western coast of Central America to the Windward Islands and to the dry forests of Brazil. Several examples of such dry epiphytic habitats from around the world, along with several examples of the more conventional extreme deserts such as the Sahara, Namib, and Atacama will be discussed in this chapter. My choices of examples have been driven by the availability of data, personal experience, and this book's emphasis on extreme environments as showcases of extreme adaptations. I have chosen the arbitrary criterion of less than 100 mm of yearly precipitation to be a necessary but insufficient condition of the conventional extreme deserts, and I have chosen examples of the epiphytic habitats based mostly on the nature of the adaptations they bring to light. The criterion of 100 mm of precipitation a year precludes discussing most of the North American deserts. (In fact, the environs of Las Vegas barely squeak in at 99.1 mm of average annual precipitation.) Only isolated pieces of the Mojave Desert and larger tracts of the Sonoran Desert fall within this constraint.

As we shall soon see, rainfall can be a poor indicator of the severity of a desert, since there are other factors that modify the availability of water; however, if enough rain does fall, all other sins are forgiven and the land becomes lush and alive. The other important factors are the following: (1) the distribution of precipitation among the seasons of the year, (2) the relative humidity of the air, (3) the amount of cloud cover (which is normally correlated with the relative humidity), (4) the wind velocity, and (5) the hydraulic properties of the local soils. The first four factors set the tempo for evaporation, while the last factor determines the amount of water actually available to most plants. The different types of precipitation are not equivalently effective in maintaining life: rain varies in intensity and duration, frozen precipitation can be moved by wind before it has a chance to melt, and dew and fog may contribute significantly to an organism's water budget if the organism has evolved the special tools necessary to harness these sources. It is often supposed that winter precipitation is more beneficial to plant life than summer precipitation, since the former has a chance to stay around longer and there would be a greater probability of root uptake were all other factors equal. Usually, they are not. For instance, Sarcobatus, Nevada, receives more than a third of its precipitation in winter; but this water is relatively unavailable to most plants because of the cold temperatures, since this locality suffers on average 144 freezing nights in winter.

An index of aridity that would apply worldwide would be very useful, and several attempts at such an index have been made, but local factors (like the ones in Sarcobatus mentioned above) make it difficult. However, attempts have held up well in certain areas. A well-known example is the biological aridity index of Emberger derived under the auspices of UNESCO for the Mediterranean region. This index is calculated by first subtracting the number of days with rain and half the number of days with fog or dew from 365, then the remaining days are weighted by a factor H that depends on the relative humidity for that day. If the relative humidity is less than 40% this factor is 1.0; at 40 to 60%, H is 0.9; at 60 to 80%, H is 0.8; at 80 to 90%, H is 0.7, and at 90 to 100%, H is 0.6. Emberger went on to create categories of biological aridity using this index. He stated that desert climates begin with an index value of 300 and absolute deserts have a value of 355 or higher. This index suffers from not quantifying the amount of precipitation received, and for regions that receive most of their moisture as fog this index can underestimate aridity; on the other hand, other indexes can overestimate aridity by not taking fog and dew into consideration.

Limiting Factors in the Desert

Whether a desert is sparsely populated by organisms or completely devoid of visible signs of life depends on the physiological availability of water to those organisms, as well as on stresses imposed by other unrelated factors (such as freezing temperatures), which in deserts take on a significance out of proportion to their severity. For instance, whether we use a biologically significant dryness index such as the biological aridity index of Emberger, or simply take pan evaporation as a measure of the transpirational stress on a plant, we find that temperate deserts such as the Gobi and the Takla Makan impose a smaller transpirational burden on the organisms living there. The same is true for many areas of Australia. However, a visit to these places shows that, at least in certain areas, evidence of life is hard to come by. Each of these deserts has a different factor that conspires with aridity to magnify its inhospitableness. In the central Asian deserts, winter temperatures as severe as those in some places in the Arctic must be withstood without the aid of a thick blanket of snow under which animals and plants could find protection. As we will see, strong insolation during winter can be extremely damaging to plants, and the midlatitude location of these deserts allows them to receive substantial radiation inputs in winter. Another conspiring factor in many deserts is the substrate. The effects of a harsh soil environment will be treated in depth in a subsequent chapter, but here we must at least discuss the essential qualities that make some desert soils extremely inhospitable.

Desert Soils

Certainly, if the concentration of nutrients in a desert's soil is low, it will support less vegetation than it would if it were soil rich in nutrients. In an ancient land like Australia, where for millions of years rain forests stood where there are now deserts, the rains that allowed those forests to grow also leached most of the nutrients out of the upper layers of the soil. As the continent drifted north, and these areas became more and more arid, the low level of nutrients exacerbated the effects of aridity and produced stresses that overwhelmed most plant lineages. The sparse plant cover allowed the surface material to be picked up by the wind and concentrated in natural topographic traps where dune fields formed. Plants have a difficult time colonizing active dunes because of the constant fight against burial on the one hand and exposure on the other. Some other areas that cannot be climatically classified as extreme deserts are almost devoid of vegetation because of the intrinsically poor moisture-holding capacity of the substrate. This effect can be clearly seen in soils derived from dolomite in parts of the Negev desert. Finally, the soil may contain toxic substances that prevent all but the most tolerant plants from growing at a locality. This sparsity of plant cover on these toxic soils translates into a lack of plant-derived organic matter in the soil, which greatly diminishes its potential water-holding capacity. This effect can be easily seen in the New Idria hills of San Benito County, California. In this area, toxic serpentine soils produce a biological "desert" surrounded by open pine forests on adjacent but more benign soils.

Another important factor in determining the sparsity of a desert's vegetation is not a physical factor, but has to do with the prospective colonists. Some lineages of both plants and animals have had to deal with deserts for a very long time, whereas others have been introduced to these environments relatively recently. The old lineages are by now well adapted to the deserts they evolved in, but adaptation to extremely harsh environments tends to be an extremely conservative process. This process produces lineages that survive the harsh physical conditions brilliantly, but are seriously threatened by milder conditions under which most other organisms produce their best growth.

Biological Adaptations

Succulence

There are numerous plant families that have stumbled on succulence as a means of storing water in order to survive hard times, but I hasten to add that succulence is an extremely varied trait that in some circumstances may not even be directly related to lack of water. The argument has been made that succulence has evolved as a response to drought in warm climates and as a response to some other environmental variable in cold climates. The argument further states that the succulence that evolved in warm climates is incompatible with a high degree of winter hardiness.

Leaf succulence in particular can be found in cold and wet environments. For example, succulent saxifrages many times grow in or close to streams where the root system will be bathed by cold and possibly anoxic water, conditions under which the roots do not function. (The plant will eventually be able to use its root system, but the most propitious period for photosynthesis will have passed.) This seems to suggest that, in the saxifrage family, leaf succulence is probably related to the difficulties of translocating water through chilled roots. No one, to my knowledge, has tested this explanation for the succulence found in saxifrages, but such an investigation would perhaps yield an unifying principle behind all succulence: the proposition that succulence evolves as a response to the physiological unavailability of water at the time of greatest need, regardless of whether the environment has plenty of water at other times. Succulence is also found in numerous salt meadow plants such as chenopods and sand-spurreys (*Spergularia* spp.), where again it seems to be related to the physiological unavailability of water at the time of greatest need. In this case, rapid transpiration can lead to an increase in the salt concentration around the roots, and as we will see in a subsequent chapter, the decreased water potential of the soil solution leads to a slowdown in transpiration when the plant requires the cooling effect of transpiration the most. This is probably only a partial explanation for the presence of succulence, since some plants, such as some of the lewisias, have succulent leaves in perfectly well-watered and relatively warm environments. In the next few paragraphs we will discuss in greater detail the types of environments where succulence appears to be of benefit, and whether each of these environments sheds some light on the cold-hardiness issue.

Another setting where succulence as a drought avoidance strategy is relatively common is that of sandy beaches. As we will see in chapter 4, the sandy coast environment is harsh, but the harshness derives from the same factors whether the organism lives on the coasts of Australia or those of Florida. This predictable harshness, together with the proximity of the sea with its long-range currents, has allowed the plants that colonize this environment to become cosmopolitan. For example, the only iceplants (Family Mesembryanthemaceae) that are found outside of southern Africa are the coastal species such as the Hottentot fig (*Carpobrotus edulis*) and the crystaline iceplant (*Mesembryanthemum crystallinum*). These plants are only circumtropical, since their frost sensitivity prevents them from spreading farther north. These two plants are particularly good examples for us because they are unabashed leaf succulents (like most of their family) and provide evidence for the argument that succulence is incompatible with cold hardiness. However, evidence against this argument can be found in a couple of members of the same family that have become common ornamental succulents in temperate countries. The hardiest of these succulents is the rosy iceplant (*Delosperma cooperi*), which does not mind winter cold down to about -20°C. This is quite remarkable when one considers that frosts are rare in its land of origin. The handful of cold-seashore succulents, as well as the very few cold-steppe and semidesert succulents, also provide evidence against the above-mentioned argument. Nevertheless, there is an undeniable dichotomy between the succulence encountered in warm environments and that encountered on cold seashores, and there are few families that have succulent members in both environments.

The presence of leaf succulents on cold shores is the best evidence that succulence and cold hardiness are not necessarily incompatible. However, the succulent plants of cold seashores are relatively few in number, with one or two species per genus and only one or two genera per family. The most notable of these seashore plants in North America are the sea lungwort (*Mertensia maritima*) in the borage family, the sea-milkwort (*Glaux maritima*) in the primrose family, scurvy grasses (*Cochlearia*) and whitlow-grasses (*Draba*) in the mustard family, and the sea-beach sandwort (*Honkenya peploides*) in the pink family. In Asia there are several seashore members of the genus *Orostachys*; they are tight little succulent rosette plants in the stonecrop family (Crassulaceae). This genus also has members in the steppes of Siberia and Mongolia. The most cold-hardy *Orostachys* close their rosettes in preparation for winter, letting their outer leaves desiccate and form a papery covering that protects the rosette. Even though in this state these plants withstand very cold temperatures, repeated freeze-thaw cycles will kill them.

All the plants mentioned up to now are perennial, since annual succulents sidestep the supposed incompatibility between succulence and cold hardiness altogether. Most of the succulent annuals of cold regions belong to the pink and goosefoot families. It is noteworthy in this regard that the plants of the purslane family—with succulent perennial members at home in hot deserts as well as temperate mountains—seem to deal with the problem imposed by cold either by becoming annuals, as is the case with the miner's lettuce (*Claytonia perfoliata*), or by becoming ephemeral geophytes, as do the lewisias (*Lewisia* spp.) of North America. In the stonecrop family, the degree of succulence certainly diminishes poleward. In the genus *Rhodiola*, the sole arctic genus of this family, most species produce annual stems from a succulent caudex. In fact, *Rhodiola* is a complex of species some of which are adapted to cold streambanks, where a small amount of leaf succulence might be of benefit. Another family that manages succulence in hot deserts and in the cold interior of northern continents is the daisy family (Asteraceae). In this family, the northern succulents are salt-plain plants (for example, the marsh elders in the genus *Iva*) that shed their branches as the first frosts arrive. For these plants, having succulent leaves in these salty environments is clearly adaptive, but they have not found a way to protect their leaves against frost.

The succulence-cold-hardiness dichotomy is most clearly evident in the borage family; the sea lungwort is one of the most successful leaf succulents of cold seashores, but in North America this species is not found south of the coast of British Columbia. Another leaf succulent in the same family is the seaside heliotrope (*Heliotropium curassavicum)*. In the New World the center of *Heliotropium* distribution is South America, but a few species have managed to march northward. *Heliotropium curassavicum* has made it to the Canadian prairie provinces, where it can be found in the many brackish marshes that dot the center of the North American continent. On its march north this plant encountered frosts that killed its aboveground parts; its roots persisted, however. It may eventually stumble upon the secret so closely held by its relative, and it too will belong to a most exclusive club.

Thus we see that many of the succulents found in the cold heart of continents and on cold coasts discard their succulent leaves in response to winter. The few exceptions—like *Honkenya peploides*, *Mertensia maritima*, *Glaux maritima*, *Plantago eriopoda*, and the several species of *Orostachys*—give us a glimpse at a remarkably rare structure in the plant kingdom: the cold-hardy succulent leaf. The mechanisms plants use to prevent freezing to death are varied and will be treated in depth in chapter 8. Except for the plantains, the other plants mentioned here have not attracted much scientific attention; therefore, their particular adaptations for survival during periods of subfreezing temperatures are not known.

If we broadly define succulence as the possession of water storage tissues, then there are several other forms of succulence besides leaf succulence; one of these is stem succulence. This form is found in some species of several different families with members native to paleodeserts; they include the cactus (Cactaceae), grape (Vitaceae), torchwood (Burseraceae), dogbane (Apocynaceae), spurge (Euphorbiaceae), stonecrop (Crassulaceae), and legume (Fabaceae) families. Some of these families produce true giant-trunked (sarcocaulescent) trees, while others produce squat treelike caricatures. The first family mentioned (the cactus family) evolved stem-succulent members in the New World only; the other families evolved stem succulents in both the Old and the New worlds. The spurge family deserves special mention, since in Africa the genera *Euphorbia* and *Monadenium* have produced dozens of species of succulents that have morphologically converged with the cacti of the Americas. But in the Americas the succulent euphorbs are few and consist of a few *Euphorbia*, *Pedilanthus*, and *Jatropha* species with swollen bases and storage roots, totally unlike the cacti with which they share the American deserts. In the Americas there are a few native cactuslike exceptions like the candelillas. (In Texas this common name is given to *Euphorbia antisyphilitica*, while in Baja California it is applied to *Pedilanthus macrocarpus*.) Judging from the vastly larger number of leaf succulents, it seems that developing stem succulence is evolutionarily more difficult than developing the several different varieties of leaf succulence. Table 1 takes a broad view of succulence and includes temperate families not normally thought of as containing succulent members. It also shows the most cold-hardy succulent genus of each family.

Table 1 Summary of Cold-Hardy Succulent Families

Family	Characteristic genera	Type of Succulence	Habitats
Agavaceae	*Agave, Nolina Polianthes, Yucca**	l,s	d
Aizoaceae	*Sesuvium**	l,s	b
Boraginaceae	*Mertensia**, *Tournefolia,*	l	b
Brassicaceae	*Cochlearia, Draba**	l	b
Cactaceae	*Corypantha, Opuntia**, *Teprocactus*	s	d,df,g
Caryophyllaceae	*Honkenya**, *Sagina, Spergularia*	l	b
Chenopodiaceae	*Salicornia**, *Salsola, Suaeda*	l,s	s,sm
Commelinaceae	*Corilla, Cyanotis, Tradescantia**	l	f
Asteraceae	*Iva**, *Senecio*	l,s	b,d,g,sm
Convolvulaceae	*Convolvulus**, *Ipomea*	l,c,r	b
Crassulaceae	*Crassula, Dudleya, Rhodiola**, *Sedum*	l,s	b,f
Cucurbitaceae	*Ibervillea, Marah, Melothria**	l,c,r	df,f
Dioscoreaceae	*Dioscorea**	c,r	df,f
Euphorbiaceae	*Euphorbia**, *Jatropha, Pedilanthus*	s,c,r	d,df,g
Liliaceae	*Aloe, Scilla**	l,s,r	g
Mesem-brianthemaceae	*Carpobrotus, Delosperma**	l,r	b,d,g
Portulacaceae	*Claytonia, Lewisia**, *Montia, Portulaca, Talinum*	l,s,r	f,m
Primulaceae	*Glaux**	l	b,sm
Vitaceae	*Cissus**	l,s,r	f,g

* denotes the most cold-hardy genus. Succulence is classified in terms of the organ(s) involved: l-leaves, s-stem, c-caudex, r-roots. Typical habitats are: b-beach, d-desert, df-dry forests, f-forest, g-grassland, m-mountain, s-salt desert, sm-salt marsh.

Even when rainfall is scant, if it is reliably concentrated in one or two seasons, an appropriate strategy of succulent plants is to hoard water in storage organs until the next rainy season comes. This approach works, precisely because of the certainty that the rains will come to the rescue before the plant exhausts its internal water supply. Such a strategy can be stretched to include regions where the cycles of wet and dry encompass more than a single year, in which case the storage organ must be quite large. Of course, for the storage strategy to work, several other adaptations must take place. The plant must be able to control its transpiration rate based on external cues, as well as on the cues coming from its internal store of water. Clearly it is only a half measure to store a large quantity of water after a rain and then transpire it all before the next rain occurs. If the plant is to survive as a viable population, other adaptations must occur in addition to a control of transpiration.

First, transpiration is one of the most important mechanisms for cooling of leaf tissues, a succulent plant that shuts down transpiration because of drought stress must be able to withstand large temperature increases of its tissues. Cacti in particular are quite capable of withstanding tissue temperatures of 60°C or above, and in fact, of the few species that have been tested, one species of opuntia is able to withstand 65°C. Second, closing down the stomata (the pores in the leaves of most plants) that allow transpiration to occur prevents the uptake of carbon dioxide, which the plant needs to make sugars; therefore, water conservation can lead to starvation. Succulents must deal with both problems in order to survive. Cacti as well as other succulents have dealt with the carbon dioxide uptake problem by inventing a mechanism to circumvent it.

CAM and C_4 versus C_3

Before we can understand the modifications to CO_2 uptake found in succulents, we must discuss the other mechanisms of CO_2 uptake found in plants. In most plants, the driving force behind CO_2 uptake is its reduction into sugars during photosynthesis. This process, which is driven by light energy, maintains a CO_2 gradient that favors CO_2 diffusion into the leaves. These plants can only fix CO_2 from the atmosphere in daylight and only by the photosynthesizing layer of leaf cells. These plants are known as C_3 plants, because the CO_2 is first combined with ribulose diphosphate to produce a molecule of phosphoglyceric acid which eventually releases the three-carbon compound glyceraldehyde. Unlike C_3 plants, many grasses, as well as other plants of hot dry environments, spatially separate the act of fixing CO_2 from the act of incorporating it into sugars. In these plants a new enzyme known as phosphoenolpyruvate carboxylase is able to incorporate

CO_2 into phosphoenolpyruvate (PEP) without the driving force of photosynthesis. The driving force for this reaction comes from the high energy state of the PEP molecule itself. The four-carbon organic acids (malic and aspartic) produced by the PEP reactions are then shuttled to other cells, where photosynthesis actually takes place. These plants are known as C_4 plants. This spatial separation allows the rate of CO_2 uptake to be fairly independent of the rate of photosynthesis, and it allows the plant to take up more CO_2 than the amount that can be immediately handled by the photosynthesis apparatus.

If a spatial separation of CO_2 capture and fixation is of benefit, then a temporal separation would also seem a reasonable approach in an arid environment. With this approach, a plant would open its stomata only in the evening, when the saturation vapor deficit of the air is at a minimum. In this way a plant would lose the least amount of water possible while taking up CO_2. Such an approach is in fact utilized by many families of desert plants. Like C_4 photosynthesis, this approach utilizes phosphoenolpyruvate carboxylase to produce oxaloacetate from CO_2 and PEP. The oxaloacetate is then reduced to malate and aspartate (some of it going into citrate and isocitrate also). For this part of the reaction to proceed, reducing equivalents are required. The CO_2 that is captured by this reaction is eventually released from these organic acids at the site where it will be turned into sugars. No matter what method a plant uses, fixing CO_2 is thermodynamically difficult, and this reaction requires a source of energy. In plants that use the PEP pathways, this energy is provided by the ATP (adenosine triphosphate) that is used to create PEP from pyruvate. A CO_2 capture, shuttle, and release system powered by ATP allows the cacti to harvest CO_2 at night, when the driving force for transpiration is lowest; with a little expenditure in energy, they gain a great savings in a much more precious commodity: water. They also release ADP (adenosine diphosphate) and AMP (adenosine monophosphate), which are necessary intermediates in many biochemical reactions. The type of CO_2 capture practiced by cacti is known as *crassulacean acid metabolism (CAM)* because these metabolic pathways were first worked out in the stonecrop family. Since then it has been found in over 19 families, most of them succulents. Succulence and CAM seem to go hand in hand, but not all succulents have CAM and not all CAM plants are succulent; the latter are fairly rare exceptions to the rule, however. These nonsucculent exceptions are the quillworts (*Isoetes* sp.), seedless vascular plants that we will encounter later in this book. These plants live in temporary pools where the competition for dissolved CO_2 is fierce.

Besides those outlined above, there are other mechanisms available to some plants that allow them to retain a positive water balance by tapping unlikely sources of water. A few trees, for example, have evolved to act as living interceptors of fog droplets. Some tree shapes are much better at fog interception than others, as anyone that has stood under a pine tree during a thick fog can attest. The Canary Island pine (*Pinus canariensis*) has evolved to depend on its ability to intercept fog moisture for a substantial portion of its water supply. We will discuss other more exotic mechanisms when we visit the places where these mechanisms come into play.

Animals in the Desert

Animals have a great advantage over plants; they can move long distances in a chosen direction and thus avoid water stress by migrating. However, small animals cannot migrate as far as large ones; they cope by finding a microclimate that is less harsh than the surface of the soil, and by taking as much advantage of this microclimate as their behavioral repertoire and physiology will allow. This type of escape allows desert scorpions to survive desert extremes mainly by avoiding them. Just as plankton move vertically in a lake, following the daily cycle of light and dark, soil-dwelling organisms in deserts migrate vertically; but in their case they are responding to the water vapor and the temperature of the soil.

There are two overwhelming imperatives in hot deserts: keeping cool and conserving water. For many animals these are conflicting requirements, and the best-adapted desert animals have found ways to keep cool while losing little or no water. The usual cutaneous perspiration experienced by most mammals is extremely wasteful of water, and desert-adapted mammals opt for other more frugal approaches. The Arabian camel (*Camelus dromedarius*), for instance, lets its core body temperature rise 6°C during the day (without triggering sweating) and cool the same amount at night. In this way the camel's own heat capacity allows it to avoid water loss. As the core body temperature rises, the camel must institute a countercurrent cooling mechanism that protects the brain from the increased body temperature. Such countercurrent mechanisms are commonplace in desert-dwelling animals, but their purpose varies. In antelopes and other large mammals, venous blood that has cooled by flowing through the nasal membranes pools in the venous sinus. Inside this sinus lies a meshwork of capillaries known as the carotid rete mirabile. The cool venous blood picks up heat from the warm arterial blood flowing through this mesh on its way to the brain. In this way the arterial supply to the brain can be cooled by several degrees. This type of countercurrent mechanism has also been demonstrated in goats, sheep and the South American rhea (a flightless bird related to ostriches).

The camel also has other ways of adapting to the desert; one of them is its ability to lose more than 30% of its body water while maintaining its blood pressure and plasma volume at reasonable levels. In comparison, a nondesert mammal suffers vascular collapse if the volume of plasma drops more than 10%. This collapse is the result of a rapid increase in the viscosity of the blood as the volume of plasma diminishes. In turn, the viscosity increase is due to the stacking of red blood cells.

Like all mammalian kidneys, the camel's kidneys can produce a hypertonic urine that allows the animal to use less water in the excretion of waste products. However, the camel, as well as other large mammals, does not seem to be able to concentrate urine to the extent that small mammals do and must compensate for it. These large mammals compensate by simply filtering a smaller volume of blood through the kidneys (the so-called glomerular filtration), thus producing a smaller amount of urine.

For most desert animals, avoiding the midday heat becomes an overriding concern, and adaptations are geared toward allowing activity during more benign times of the day. A case in point is the roadrunner (*Geococcyx californianus*), perhaps the most characteristic bird of desert lands of the American West. It is one of the few cuckoos that work for a living. It tends its own young, feeding them mostly desert arthropods, snakes, and rodents. The roadrunner has a patch of black skin on its back that can be exposed by lifting the back feathers. This is an ingenious device designed to allow the bird to come up to speed early in the morning, while its prey is still sluggish from the previous night's chill.

Dew Harvesters

The smaller an animal is, the larger its surface to volume ratio and the larger its loss of water per unit of weight; however, small desert animals have developed behavioral adaptations that allow them to avoid the worst of the desert environment. For example, many rodents take advantage of the cooler temperature inside their burrows. Inside the burrow, the temperature can be more than 30°C cooler than the surface air temperature and the humidity can be 80% versus 25% at the surface. One Middle Eastern rodent even has the remarkable habit of building a tiny pile made of small pebbles just outside its burrow's entrance. These small rocks have a large surface area and cool down fast as evening comes (mainly by radiation to the night sky). As the warmer air of the burrow slowly exits through the entrance to make way for the cooler air now ponding at the surface of the ground, the moisture in the warm air condenses on the little pebble pyramid. The rodent drinks the dew that has collected overnight and thus recovers some of its own exhaled moisture. Many other examples exist of behavioral adaptations that allow mammals to harvest desert dew. Even large grazing mammals may benefit from such dew-harvesting strategies. In the Kalahari desert (a semidesert despite its name), the herds of eland (*Taurotragus oryx*) and springbok (*Antidorcas marsupialis*) forage at night, when the dew collects on the grass increasing its moisture content to 30% from 1% during the day.

Arthropods, however, have devised the largest number of behavioral adaptations for the harvesting of dew. Several species of darkling beetles (Tenebrionidae) of the Namib Desert have developed behaviors that allow them to harvest the fog that rolls in from the ocean. One beetle species digs a trench near the crest of a sand dune and parallel to it. This trench collects the small amount of water that moves downhill along the surface of the sand. This technique is the same as that used by ancient human inhabitants of the Negev, as well as of other areas in the Middle East. These ancient cultures used the channeled water to irrigate their crops and also stored some of it in cisterns, something the beetles would be hard-pressed to do. Another darkling beetle positions itself into the wind and lifts its abdomen so as to expose the most surface for the condensation of the wind-borne moisture. As the water collects on the beetle's elytra, it runs to the beetle's underside and makes its way toward its mouth. Some spiders use their webs as moisture traps. Orb weavers normally place sticky globules at regular intervals along their web as a means of entrapping insects. Some orb-weavers, such as the common European garden spider (*Araneus diadematus*) have adapted this behavior for dew harvesting. The beads these spiders secrete are particularly hygroscopic. Sometime during the night, as the air temperature drops below the dew point, these beads are laden with water. In the early morning, the spider simply eats the web to collect this harvest of moisture.

The Kidney

The unsurmountable problem small size produces in animals is the inability to cover large distances in reasonable time. Whereas a large mammal has a certain efficiency of scale that allows it to cross significant stretches of barren desert between its watering area and its feeding area, a small mammal is constrained to travel shorter distances at night and during crepuscular times. In fact, the territory of some small desert rodents can be as small as several square meters. This size handicap forces animals to choose between areas with food and areas with water. Those mammals that have chosen areas with food and have sacrificed proximity to water have embarked on a difficult road, but some of these creatures have succeeded admirably. Rodents like the kangaroo rats (family Heteromyidae) in the New World and the jerboas (family Dipodidae) and Libyan jirds (family Cricetidae) in the Old World do not need to drink liquid water; they can obtain enough moisture from the seeds and plants they consume (although they may not thrive on such small water rations). How can animals feeding on material that is only 5% moisture obtain enough water? The answer lies in the mammalian kidney.

The kidney is an ancient organ; the basic ultrafiltration unit is found in one guise or another in many metazoan phyla. In vertebrates, the ultrafiltration unit is called the nephron, and it has a distinct form and design that does not deviate from one group of mammals to another. What changes is the relative sizes of the components of this basic unit. Mammals took the basic plan of the nephron and elongated a seemingly insignificant portion of it. This portion, called the loop of Henle, is crucial to the functioning of the nephron in a hypertonic environment. The kidney is the

The 25 species of jerboas range from the Sahara to the steppes of central Asia. They all have the long tail and hind legs characteristic of the family.

only concentrating organ in mammals, and small mammals have taken this organ's design to its ultimate development; in the process some have managed to develop the ability to concentrate their urine up to twenty-five fold above their blood concentration. If we consider that the osmolarity of blood is about 0.9% sodium chloride equivalents, a 25-fold ratio implies a urine osmotic pressure equivalent to a 22.5% table salt solution, or about as concentrated as Utah's Great Salt Lake. Unlike mammals, amphibians and reptiles do not have an elongated loop of Henle and cannot produce hypertonic urine in their kidneys; they must rely on extrarenal recovery of water in order to survive the desert environment. Birds' kidneys have a mixture of reptilian-type and mammalian-type nephrons, and they are able to concentrate their urine several times above their blood concentration. This is not enough in extremely dry environments and some birds, like some reptiles, have developed salt-excreting glands that are normally located around the nares or the eyes.

The nephron works by a clever uptake and release of sodium ions that establishes a concentration gradient from the cortex to the medulla of the kidney and leads to the extraction of water from the lumen of the distal tubule. This process still wastes a lot of water, and it would be much more efficient to produce insoluble waste products that need very little water for excretion. This is the approach of birds and reptiles and of the best-adapted desert frogs. In these vertebrates no water need be used to excrete nitrogenous wastes which take the form of urate and guanine salts, since the solubility of these salts is extremely small; but water still needs to be used to excrete electrolytes, and it is here that salt glands come into play.

The Geological History of Deserts

Most of the current deserts on Earth have relatively recent origins, but there are some ancient ones as well. Among the former are many parts of the Sahara and of southwestern North America that had rainier climates as recently as ten thousand years ago. Among the latter are parts of the Namib of southwest Africa and of the South American coastal desert. In the Namib we find strange desert-adapted gymnosperms such as *Welwitschia mirabilis*; some individuals of this species might be one thousand years old. The age of the southern African deserts has allowed a whole family of plants to explode onto the scene, with what looks like almost infinite ingenuity and variety (2000 species strong); the iceplant family has radiated into diverse forms, from deciduous bushes to rock mimics, in less than five million years. In the South American coastal desert another family, the Nolanaceae (related to the tobacco and borage families), has developed into more than eighty species of halophytic shrubs and herbs able to withstand some of the harshest conditions in the world.

There are several well-understood factors that conspire to make a desert; some of these factors are on a planetary scale, while others are smaller-scale phenomena. The Earth is tilted 23.5 degrees on its axis; this tilt means that at some point during the year, the surface of the planet between 23.5°N latitude and 23.5°S latitude will receive direct rays from the sun. The parallel at 23.5°N latitude is known as the Tropic of Cancer, and the parallel at 23.5°S latitude is the Tropic of Capricorn; these two parallels define the tropical and subtropical areas of the globe. In these areas, the more direct rays of the sun cause more heating and a greater upward movement of air, creating a well-defined low pressure area that encircles the globe. As the air rises from the tropical regions it expands and cools, allowing water vapor to condense and fall as rain. This flow of tropical air must go somewhere, and as it reaches the upper troposphere, the flow diverges toward the poles, cooling as it goes. As this flow cools it continues losing moisture until it becomes denser than the underlying air. At this point it descends back to the surface of the Earth along bands centered at about 30°N and 30°S latitude. This falling, denser air creates a high-pressure ridge centered at these latitudes. As the air falls it warms and its water deficit increases until, by the time it reaches the surface of the Earth, conditions are ripe for the creation of deserts. This planetary-scale convection phenomenon is known as a Hadley cell, after the eighteenth-century scientist who proposed the first reasonable explanation of the trade winds, which are a surface manifestation of the cells that now bear his name. These cells also repeat from 30 to 60° latitude, and again around the poles. The sinking air at the poleward leg of each cell induces stable atmospheric conditions propitious for the development of deserts. The Baja California deserts, as well as the Sonoran Desert and the deserts of the Middle East can be partially explained by this planetary phenomenon. A similar set of cells exists in the Southern Hemisphere, and these contributed to the creation of the Australian deserts. There are other factors at play, however.

A quick look at a map will show that, even though many of the larger deserts do fall along the 30° parallels, many well-watered areas do also. Obviously, the creation of desert conditions must also involve other factors. These other factors, mainly complications of topography and ocean currents, are what makes each desert unique. One of the most important phenomena is caused by large land masses; as moisture-laden air travels over them, it loses moisture as precipitation, without significant replenishment. Therefore, the centers of continents tend to be drier than their peripheries—a factor that highly influences deserts such as the Gobi and the Takla Makan of central Asia. The Takla Makan is also influenced by the rain shadow of the Himalayas.

The orographic or rain shadow effect is important in many other areas also. It sometimes produces extremely arid valleys, as is the case in Death Valley, California, where the almost total encirclement by the Panamint, Funeral, Black, and Grapevine mountains makes for one of the most arid places in North America. The taller and more massive the mountain range, the more effective it is at wringing moisture out of the clouds. This is the reason (as we will discuss in chapter 10) that the Chang Tang and the Puna are such dry environments. Their aridity is ameliorated somewhat by the thin atmosphere, which does not retain heat well and allows these places to cool, thus diminishing evaporation. In North America, rain shadows of the Cascades and the Rockies allow semideserts to develop as far north as the inland valleys of British Columbia and Alberta. Even though places such as the Okanagan Valley receive only 310 mm of precipitation, their northerly location prevents the temperatures, and therefore the evaporation, from outpacing precipitation by a large amount.

A third mesoscale factor in the development of large deserts is the interplay of ocean currents with the land. The large ocean basins of the southern and northern hemispheres have large gyres that bring cold polar waters toward the equator. The paths these gyres take are defined by the continents that bound these basins. In the eastern Pacific the Peruvian (Humboldt) current is the equator-bound arm of the Southern Pacific Gyre. As the Humboldt current travels northward it encounters progressively warmer water. Along the coasts of Peru and Ecuador, the cold water of the current sinks, displacing the nutrient-rich bottom waters upward. This upwelling is extremely important for the ecology of South America's west coast. In the years when the Humboldt current moves too far offshore, the local supply of nutrients vanishes and as a result fish become scarce. This event is known as El Niño, since it happens around Christmas.

The effect of the Humboldt current on the climate of the Pacific coast of South America is profound. As the air travels over the cold water before reaching the land, it cools and loses a large part of its moisture, robbing the land of much needed rain. But the cool air still contains some moisture, and in many areas heavy fogs are common during the winter months. Ironically, the warm water that accompanies the El Niño event produces substantial but unpredictable rainfall in areas that may go several years without a drop of rain. The same scenario repeats itself along the southwestern African coast and again along the southwestern Australian coast, but in these regions, the climatic oscillations due to changes in their respective cold currents are not as pronounced as El Niño.

The constant motion of the continents has made deserts appear and disappear over spans of millions of years, and oscillations in climate have had the same effect over spans of tens of thousands of years. There is substantial geologic evidence that the central Asian deserts were lush forests 30 million years ago. Sixteen million years ago, the same area was mostly occupied by a large inland sea, the remnants of which persist as the Caspian Sea. By then, Africa had settled close to its present location, but Australia continued to move toward the equator. At the same time Patagonia was close to 30°S latitude and held a large inland desert (parts of Patagonia are classified as semidesert even today). The progression of the continents has meant that some deserts had a geologically transitory existence, while others have existed for many million of years.

Biologically there is a vast difference between old deserts and infant deserts such as most of those in North America. The harshness of a desert environment and the relatively small number of individuals of a species that can eke out a living in it prevent evolution from proceeding at a fast pace. The raw material for adaptation is mutation; the harsher an environment is, the likelier it is that any mutation will be detrimental to the reproductive success of the individual. On the other hand, if an environment becomes harsher over millions of years, this very slow deterioration allows mutations that at first might be detrimental to combine into a useful adaptation. If harsh conditions persist for millions of years, then these adaptations will be fine-tuned to such an extent that, were the area to return to milder conditions, the organisms possessing these adaptations would find themselves unable to compete or to defend themselves against predators or parasites. For example, many desert plants are exquisitely susceptible to high soil moisture and subsequent fungal attack.

The slow movement of the African continent and its position away from the centers of ice age activity have allowed certain parts of the Namib to remain a desert for at least 20 million years. The same can be said for a part of what is now the Sonoran Desert, a part of the western Australian desert, and a part of the South American coastal desert. In all these regions strange forms—unique plants that have no temperate or even humid tropical relatives—have developed. Families such as the Nolana (Nolanaceae), ocotillo (Fouqueraceae), iceplant (Mesembryanthemaceae), welwitschia (Welwitschiaceae), and jojoba (Simmondsiaceae) are true desert natives. The places they call home are the topics of the next section.

A Tour of Extreme Deserts

The Aerial Deserts

The bromeliads are the prototypical epiphytes, growing on any and all trees of the neotropical wet forests. There are tank bromeliads that weigh hundreds of kilos burdening down branches that will eventually break, taking the bromeliads into the dark recesses of the forest where the light is not strong enough to allow vigorous growth. Such a plant will hang on for at least a while, long enough for some forest traveler to marvel that a massive rosette the size of a respectable agave survived dozens of meters above the forest floor simply on the rain it collected in its central tank. During the day these little ponds look like any other water-filled cavity, but at night life stirs within them. Once, while looking for frogs in the Costa Rican rain forest, I was startled by a large wolf spider. Following this spider over a large bromeliad I lost it amongst the central leaves, only to catch a glimpse of something retreating into the central water-filled tank. Flashlight in hand I leaned over the top of the bromeliad to look at what had retreated into the water and was now hidden by the dead leaves at the bottom of the central tank. To my amazement and delight it was a large tarantula, its abdomen the size of a large thumb. I proceeded to investigate all of the bromeliads I could find and discovered that almost all the large ones were occupied by resident spiders, each roughly proportional to the tank it inhabited. I then quietly hunkered down and waited for the frogs I could hear all around to appear.

The tank approach works well in wet forests, but in seasonally dry forests the epiphytes must have other means of dealing with the lack of water. Of the bromeliads, the best adapted for a seasonally dry existence are the tillandsias and their relatives. Tillandsias have an amazing organ known as the absorbing trichome. Trichomes are not rare in the plant kingdom, but tillandsias have taken the structure to a new level of complexity, and possibly a new purpose altogether. These structures serve as one-way valves in the uptake of liquid water falling in the form of rain, fog, or dew. We will see how this adaptation has allowed the tillandsias to thrive in the coastal deserts of Peru and Chile where the only significant source of moisture is the coastal fog. But first we will explore the mechanism of this valve. Figure 1 shows a stylized schematic of a tillandsioid trichome in both the closed and the open state. The shield part is composed of the cell walls of formerly living cells. The original parent cell divided to form a neatly geometric pattern of a central disk of four cells surrounded by an eight-cell ring in turn surrounded by a sixteen-cell ring. The living material then died leaving behind nonliving, but animated "cells." The central-disk cells have a thick outer cell wall, while the eight-cell ring has somewhat thinner outer cell walls and the sixteen-cell ring has thin outer cell walls but thick inner cell walls. The thick cell walls are not suberized and lack a cuticle. The material they are made of (mainly cellulose) is quite hydrophilic, and it draws water in through capillary action into the empty lumina of the shield cells, just as a cellulose sponge does. And just like a sponge, once the water enters the empty lumina of the cells, their walls expand. This expansion flattens the shield and opens a path for the intake of liquid water. The trichome of each species is tuned to that species' particular moisture requirements, and it seems that, at least for some species, trichomes can even be used as one-way valves to take up moisture from merely humid air.

The distinction between fog and merely humid air is biologically important, since many organisms have found ways to collect fog and dew, but few can avail themselves of water in the vapor phase. A few species of tillandsia seem able to accomplish this feat. The trichomes of these few species are able to equilibrate with the atmospheric water vapor like the pages in a book on a humid day. It has been proposed that some of the dome cells—the living cells beneath the central-disk cells—are able to produce high concentrations of osmotically active solute that can be excreted just under the central-disk cells producing a strong osmotic gradient and, just as a dusting of table salt is used to draw out water from eggplant slices before

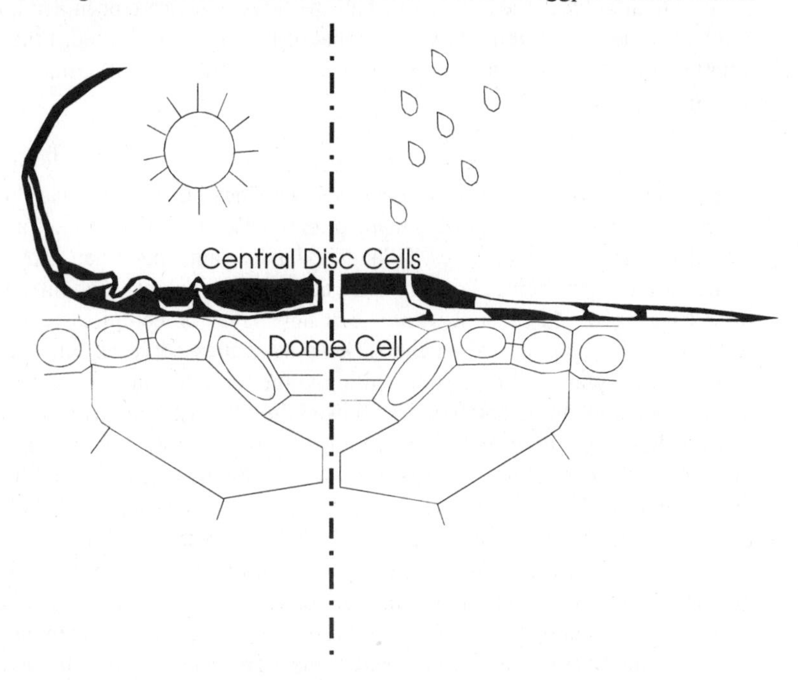

Figure 1 The tillandsioid absorbing trichome in cross-section.

cooking, this gradient will draw water from the lumen of the disk cells. The dome cells must then pick up this diluted solution and take it back into the cytoplasm where the osmotically active solute is polymerized or in some other way made to release its load of water, and the water is then free to continue its journey to the cells that need it most. This is only one of several mechanisms that have been proposed in order to explain the operation of the absorbing trichome. These proposed mechanisms produce a net water gain whenever the relative humidity is high enough to open the trichome and the internal water potential is more negative than that of the air. Obviously, if the internal water potential is higher than the external one, water vapor will flow out of the plant. The trichomes of plants that are able to sustain themselves on atmospheric water vapor must remain closed at such times. These times occur when the relative humidity of the outside air is higher than the relative humidity in equilibrium with the osmotically active solution found under the central-disk cells.

In the absorbing trichome we have a dead structure that is nonetheless able to act as both a sense organ and a mechanical valve rolled into an attractive little package. Thus, some tillandsioids have a third possible source of moisture (besides rain and fog) in merely humid air, and it has been shown that at least some tillandsias can absorb water vapor from air at 80% humidity. This remarkable water-vapor-absorbing ability will only be useful in deserts with periods of high humidity. To associate deserts with high humidity is not as contradictory as it might at first appear, and the desert islands of the Gulf of California and the Persian Gulf are examples of arid lands where the air can be almost saturated with water vapor for significant periods of time, especially at night. The water-vapor-absorbing faculty may be quite common in plants of fog deserts, such as parts of the Atacama and Namib. However, it is very difficult to determine whether plants in the wild derive a substantial part of their water budget from humid air, since humid air and fog go hand in hand. For this and other reasons, the significance of humid air in the water budget of plants has not been systematically explored.

Coastal fogs are significant and easily quantified sources of moisture in the American deserts from Baja California to central Chile. In the Baja deserts of El Vizcaíno and Llano Magdalena two species of native tillandsia grow on any available surface above the ground, including telephone wires and columnar cacti such as the cardón (a relative of the saguaro of the Arizona and northern Sonoran deserts). A relative of the tillandsias, the genus *Hechtia*, is also plentiful on the sheer walls of canyons in the El Vizcaíno desert. All these plants can be shown to be able to derive a large part of their water budget from fog or dew.

A tantalizing glimpse of tillandsias' ability to extract water in areas that seem to have none can be seen in the Atacama Desert, on the coastal plain west of the Pampa del Tamarugal. As we travel north through this area, the vegetation becomes sparser until the only plants that are left are tillandsias. In some sandy areas they are quite numerous, and even though these tillandsias are not rooted in the sand, the sand has a beneficial effect on their moisture supply. This is an example of how deep sands can provide a more stable moisture regime than the surrounding desert (we will encounter other examples in chapter 4). They retain their epiphytic form and function even without any trees! Fog forms frequently in this area, but the question remains: Why are tillandsias the only successful vascular plants in the vicinity, whereas in other fog belts a few hundred kilometers north whole forests seem to thrive on a similar ration of fog?

The Sahara

An example of the evolutionary conservativeness shown by desert inhabitants can be seen in the Sahara, which biologically is composed of two distinct zones. The flora of the southern Sahara is mainly dominated by plants of palaeotropic origin, mostly small trees and thorn shrubs. Of interest here is the almost complete lack of frost tolerance in plants of this region. This evolutionary limitation translates into a limit to their northward movement into the northern Sahara. Conversely, the northern flora is composed mainly of dwarf shrubs of holarctic origin. Composite shrubs and chenopods are richly represented in this flora. Only in a few places do the two floras actually intermix, even though their common boundary extends over 5000 km. Intermixing is prevented because much of the common boundary is so arid that no plant or animal can survive there.

In the northern Sahara, the plant lineages are from a holarctic source, and like in most other holarctic deserts, the flora is dominated by widely spaced shrubs. Trees are conspicuously absent in this region. The northern Sahara can have cold winter nights, and this seems to be the barrier that prevents the southern flora from invading the somewhat less xeric northern region. The flora of the southern Sahara is extremely well adapted to some of the most xeric conditions on the planet; it evolved from the savanna flora of the Sahel region (the band of dry grassland and thorn forest that bounds the Sahara to the south and is now turning into desert through overutilization). However, the vegetation in the southern Sahara is limited to trees and tree-like shrubs that occupy the dry creek beds or wadis, and the rest of the landscape is devoid of plant cover. Areas in this region that are apparently suitable for the northern shrub flora are completely devoid of vegetation. It has been suggested that the somewhat greater aridity in the south is what prevents the northern bushes and even the native southern trees from invading these barren areas.

Areas of tension between desert floras of different lineages are not confined to the northern hemisphere. For instance, along the northern border of the Succulent Karoo Region of the Namib Desert, a palaeotropic flora similar to that of the southern Sahara meets the dwarf succulent flora of the South African deserts dominated by the iceplant family, one of the preeminent succulent families.

A similar scenario exists in North America, where the boundary between the northern Great Basin semidesert and the southern Sonoran Desert takes on unique biological attributes and its own name, the Mojave Desert. In these deserts the northern lineages are, again, shrubs of holarctic origin; but the southern flora has a distinct sarcocaulescent component, together with a variety of small leguminous trees (many with thorns), and large columnar cacti.

The Negev

As rain becomes more and more infrequent, the actual amount of precipitation becomes much less important than the fate of this water in the soil. Small showers in a desert like the Negev, located mainly in southern Israel, tend to evaporate and contribute only marginally to the water supply available to plants. L. Shanan has shown that as the rainfall amount increases so does the percentage of the rain that actually penetrates the soil. Whereas a rain of 10 mm may have only 15% penetration, one of 90 mm may have 30%, so that the water actually available to plants is 1.5 mm and 27 mm respectively. Almost nothing can grow on the former, but the latter is a respectable yearly allowance for a desert plant. Plants growing in the Negev must also contend with highly unpredictable rainfall events. One of the best adaptations to highly irregular and scant rainfall is the "survival by partial death" syndrome. In this approach a plant gives up photosynthetic tissue gradually, starting with the leaves and progressing to the small twigs and then the larger branches. This is the approach taken by the dwarf shrubs of the genus *Zygophyllum* in the caltrop family (Zygophyllaceae). They are some of the most drought-hardy perennial shrubs of the Middle East, and with seventy or so species, dominate severe deserts throughout Asia, Africa, and Australia. Most *Zygophyllum* species possess leaves that are small and succulent, and in times of drought are the first organs to be discarded. If the drought persists the twigs are shed, and if the drought lasts for several years, large parts of a single plant might die.

In many woody plants, different parts of the root system supply water and nutrients to different parts of the canopy. The roots of a desert plant can travel for long distances in order to reach cavities in rocks or deep sands where water might escape the unrelenting imperative to evaporate. Sometimes a large rock can act as a concentrator of rain, allowing whatever rain falls on it to run off quickly and concentrate on the shady underside where it has a chance to percolate. One part of the root system of a desert plant may be under a large boulder, while another part may not have been so lucky and extracts whatever moisture it can from the soil under a small depression. The *Zygophyllum* plant survives drought by letting die those roots that are having trouble extracting moisture and the aboveground parts that depend on them.

In the Negev another successful strategy is that of the perennial ephemeroid. This rather contradictory term denotes a plant that has a perennating structure below ground or partially breaking the surface of the ground, but above ground behaves as an ephemeral annual. This group of plants can be further divided into those that have only leaves and flowers above ground (geophytes) and those that produce annual aboveground stems (hemicryptophytes). One of the most successful geophytes in the Negev is the sedge *Carex pachystylis*. The most successful hemicryptophyte is *Artemisia herba-alba*. These plants grow during the winter, and during the summer they avoid the heat and drought by discarding the aboveground parts completely. This approach is quite common among *Carex* and *Artemisia* species faced with adverse conditions. In fact, the same strategy has been adapted by the two shrubby artemisias that grow in Alaska, but for a completely different purpose. As a means of surviving bitterly cold winters, these plants discard the year's branches and only retain the perennating woody base. This type of growth might seem to be a clear example of one step forward and two steps back, until one realizes that most of the biomass of these plants (geophytes and hemicryptophytes) is under the ground, and this part grows whenever there is enough moisture.

Lichen Feeders of the Negev

Just as biodiversity increases as rainfall increases, so does the biodiversity of extreme deserts increase as the amount of dew deposition or fog interception increases. When moisture does not actually condense, but humidity is still high, lichen diversity can be quite high at the expense of vascular-plant diversity. A high lichen diversity and standing crop attracts consumers just as vascular plants would. Some of those consumers are desert snails.

Lichen feeders abound in certain deserts. In the dolomite barrens of the Negev, three closely related species of small (1-cm long) snails—*Euchondrus albulus*, *E. desertorum*, and *E. ramonensis*—fulfill this role. They scrape the rocks on which they live for sustenance. When they feed, they tilt their shells to a more upright position, and with a swaying back-and-forth motion scrape a groove 1 mm wide by 0.5 mm deep. They are not particularly well equipped to subsist on lichens, and out of all the rock and lichen material they ingest, they only digest 5%.

Lichens can grow in these dolomite barrens because from September to April, for an average of 210 days out of the year, dew collects on these rocks. If we apply these numbers to the biological aridity index, this place barely qualifies as a desert, a clear misrepresentation. The biological aridity index emphasizes dew and fog to the same extent, while in reality fog can supply a larger amount of moisture, whereas dew is a more limited, even if more common, phenomenon. The 30 mm of dew condensation per year complements the meager rainfall of 100 mm per year to make this particular area a marginally habitable one.

The snails' working day starts at sunset. They first defecate their previous night's meal before venturing out from their under-a-rock shelter onto the surface of their rock islands. After scraping the rock the whole night long, they then retire at sunrise. In summer there is no dew, so the snails aestivate by closing themselves within their shells. When the rainy season returns, the ground becomes wet enough to allow individuals to cross the areas separating the rocks and visit their neighbors, every single one of which is a prospective mate.

The population of these snails is not large and amounts to no more than 20 or so individuals per square meter. The snails may be helping the lichens grow by exposing new rock surface and thus opening up real estate for the lichens. The grooves the snails produce may also be better at collecting dew and nitrogen-rich dust than the old rock surface; in this regard their behavior is reminiscent of that of the darkling beetles of the Namib. Over a year's time the snail population grazes 47% of the total rock surface to a depth of 1 mm; that is 900 kg of feces per hectare, compared to deposition of wind-imported dust, which is estimated to bring in only 360 kg of soil per hectare. The snails' feces contain 5% nitrogen, making them a fair fertilizer; since they are deposited on the underside of the rocks, they are a more effective fertilizer than windblown soil. Dolomite barrens with similar populations of snails and lichens exist all around the Mediterranean Sea, extending from the Dolomite Alps of northeastern Italy to the Middle East.

The Sonoran Desert

The Sonoran Desert encompasses all of Baja California and most of the state of Sonora in Mexico. It is a distinct province having a distinctly tropical biota dominated by creosote bush (*Larrea tridentata*). Like the southern Sahara, it clashes to the north with a much younger desert, the biota of which is of mixed parentage with both northern and southern components. This of course is the Mojave Desert, and because the Mojave is delimited to the north mainly by the geographical extent of creosote bush, some authors simply label it as the northernmost province of the Sonoran Desert.

The creosote bush is the dominant plant in the most extreme deserts in North America. As a member of the caltrop family, it also adopts the "survival by partial death" syndrome we saw in *Zygophyllum*. Not only can it shed its leaves in case of extreme drought, but it may also shed twigs and whole branches if conditions do not improve. It is the ruler of its domain, which extends over 30 million hectares in Mexico and 15 million in the United States. No wonder a common Spanish name for it is *gobernadora* (governor). Its range is expanding mainly due to its unpalatability to cattle, which devour creosote bush's tastier competitors. It only cedes dominance where the soil is too salty, and in such instances the chenopods (especially the several species of *Atriplex*) are more than eager to colonize.

The caltrop family dominates deserts that are in a sense intermediate between cold deserts like the Gobi and warm deserts like the Namib. Many members of the family are able to withstand mild frosts, but when the temperature drops just a little bit further many are irreparably damaged (some of the *Zygophyllum* species are important exceptions to this rule). This is a satisfactory explanation for the northern extent of creosote bush in southwestern Utah and adjacent Nevada, where it can be seen dominating the hot, wide valley bottoms until one reaches its northernmost populations. At its northernmost extent, a stunted form of this bush is present at midelevations and absent from the lowest depressions, where the four-winged saltbush (*Atriplex confertifolia*) takes over. This pattern of distribution is caused by cold air ponding in these depressions in winter, making them deadly frost traps where creosote bush cannot survive.

The creosote bush is the sole North American member of its genus. Its closest relatives are found in the deserts of Argentina, especially the Monte desert of the central part of the country. There we find the jarillas (*Larrea divaricata* and *L. cuneifolia*), two of only five species in this genus (all of them American). The creosote bush and *Larrea divaricata* are so similar that until about 1965 they were thought to be one species.

Larrea species are special in several respects. Like other extremely well-adapted desert plants, they are able to orient their leaves so that the solar heating will be minimized. The most proficient at this is the compass plant (*Larrea cuneifolia*) of the Monte desert. It grows its leaves vertically along a north-south axis, thus avoiding the desiccating rays of the midday sun. The *Larrea* species also produce a sticky varnish that may help the plants diminish their transpiration.

Pollen evidence indicates that creosote bush was present in the lower Colorado River 17,000 years ago at a time when the Wisconsin Glaciation was starting to give way to a slow warming. About 10,500 years ago the warming sharply accelerated, and creosote bush was poised to conquer the land vacated by the piñon forests that occupied what is now the Mojave and northern Sonoran Deserts. The original diploid population can still be found in the Chihuahuan Desert, whereas the northern Sonoran population is tetraploid, and the Mojavean population is hexaploid. In the Nevada desert, near the California border, there are several rings of bushes that are over 25 meters in diameter. These rings are in fact made up of clones of the same original plant that are maintained through suckers. If these rings of clones expanded in the past at the same rate as in the present, we can calculate an age of 10,000 years for them. In other words, they are exact replicas of some of the original colonizers of the Mojave Desert. Even older clones exist farther south, and the oldest has been estimated to be 11,700 years old, making it the oldest multicellular organism on the planet. That no clones reclaim the centers of the rings is somewhat of a puzzle, but the battle for moisture going on underneath the ground may be a sufficient explanation for this outward march.

Even though the creosote bush is thought of as the dominant Sonoran Desert plant, other species abound and in fact may outnumber the creosote bush 9 to 1 in many situations. This is the case with the white bursage (*Ambrosia dumosa*) and several other small composite shrubs. In the driest parts of the Sonoran Desert, such as the Desierto de San Felipe of northeast Baja California and the Desierto De Altar of Sonora, creosote bush and white bursage may constitute 90% of the scant vegetation. Some of these areas receive only 30 mm of yearly precipitation, with pan evaporation rates approaching 3000 mm per year in the summer. The ratio of evaporation to precipitation is therefore over 100, even greater than that recorded at Greenland Ranch in Death Valley (a mere 32.3). There are some places on the planet that get much less rain, however.

The Atacama and the Peruvian Deserts

The west coast of South America from about the Peruvian-Ecuadorian border in the north to La Serena in the south is one of the driest places on Earth. Within this 3500-km desert coast there is a place so inhospitable and dry that after traversing any part of it the sight of a plant, any plant, becomes startling. This is the Atacama Desert of northern Chile. In some places there has been no measurable rain in decades, but the biological communities that manage to live on this inhospitable coast have evolved from the middle Miocene to rely not on rain, but on the fog that develops as the cool air moves onshore. The factors that have led to arid conditions on parts of this coast for the previous 14 million years depend on the position of the South American and Antarctic continents and the height of the Andes, factors which change slowly over the millennia. The Antarctic is responsible for the Humboldt (Peruvian) Current, which has followed the western coast of South America for the past 14 million years, ever since Antarctica settled in its polar location. Such a cold current tends to produce an inversion layer in the atmosphere, where colder air lies at the surface overridden by warmer air. Such a situation inhibits convection, and without convection clouds do not form. This maritime inversion by itself is not sufficient to produce such arid conditions, however. There must also be some factor that forces this cold surface air over the land; in the case of the western coast of South America, this factor is the persistent anticyclone that exists around 30°S latitude. In the Southern Hemisphere anticyclones rotate counterclockwise, and the flow of air is toward the South American continent at the latitudes of the Atacama Desert. The last factor that prevents any relief from these drying influences is the presence of the Andes mountains. The Andes have shaped South America in ways we have only begun to understand. In terms of the climate, the Andes are a very effective barrier to the movement of Amazonian moisture westward. The geologic evidence suggests that these conditions have prevailed over at least some parts of this coast for a very long time indeed.

Pampa del Tamarugal

The Pampa del Tamarugal is one of the driest parts of the Atacama Desert, and is almost completely devoid of life. The landscape shows the characteristic patterns of wind erosion, and one must travel back 14 million years in the stratigraphic record to find significant evidence of water erosion. This forbidding place has been conquered by an organism that in so doing has generated astonishment and admiration. This organism is the large tree *Prosopis tamarugo* (large by desert standards) that grows in monotypic stands. For some time it was thought that these trees had stumbled on some magic recipe for desert survival, and it was postulated that they somehow reacquired their transpired water as dew. It is now known that these trees, like many others in the genus *Prosopis*, are deep-rooted phreatophytes and survive on deep underground reservoirs of water.

There are other species of the genus *Prosopis* in other deserts of the world. In North American extreme deserts, such as the Desierto de San Felipe in Baja California, five species of *Prosopis* (the mesquites of North America) dominate bajadas and dry arroyos. It has even been reported that the roots of one particular Arizona mesquite were able to reach 50 meters into a mine. Halfway around the world, the last tree left in Bahrain's sandy central depression is an old specimen of *Prosopis juliflora* reverently termed the Tree of Life. This tree shares the central depression with mounds of *Zygophyllum qatarense* and *Heliotropium ramosissimum* thinly spread out over the sand sheet; these subshrubs are true desert inhabitants that may not share in the wellspring that sustains the Tree of Life.

What makes *Prosopis tamarugo* unique even among *Prosopis* is its ability to grow rapidly, as if it were a cottonwood or some other floodplain tree, by tapping water that may be 12 meters below the surface of the ground. *Prosopis tamarugo* grows in two distinct areas of the Pampa del Tamarugal: the Salar de Pintados and the Salar de Bellavista; these two salares (salt plains) have thousands of hectares of these trees. These groves, over 85% of which are planted, subsist and in fact thrive in an area with an average rainfall of under 1 mm per year. Phreatophytes, more than most desert species, depend on the infrequent occurrence of wet years to establish a presence, and when wet years go from infrequent to unheard of, phreatophytes tend to become a static relic population of old trees, as has happened to the Tree of Life.

In the case of *Prosopis tamarugo*, humans have intervened, providing this tree with a significant advantage. When left to their own devices, phreatophytes must handle such environments in one of two ways, either the sexually mature plants must be long-lived, or the seeds must be long-lived. Legumes such a *Prosopis* are already genetically equipped to produce long-lasting seeds that can only be enticed to germinate when the improved conditions have passed a set of permanence criteria. If a phreatophyte seed were to be fooled into germinating by a light shower, the seedling would be doomed. On the other hand, in a year when the El Niño produces torrential rains that flood the creek beds where the phreatophytes normally live, a slowly sinking water table will allow the seedling's taproot to keep pace. In North America, other *Prosopis* have been clocked doing a meter per year as they dig deep into the substrate seeking the retreating liquid commodity.

The Lomas Formations

We have already discussed the adaptation that allows plants and animals to subsist with only fog or dew, and now we take a look at the areas where this is the predominant form of survival. In both the Chilean and Peruvian deserts, there are hills that intercept the fog that moves inland; this happens

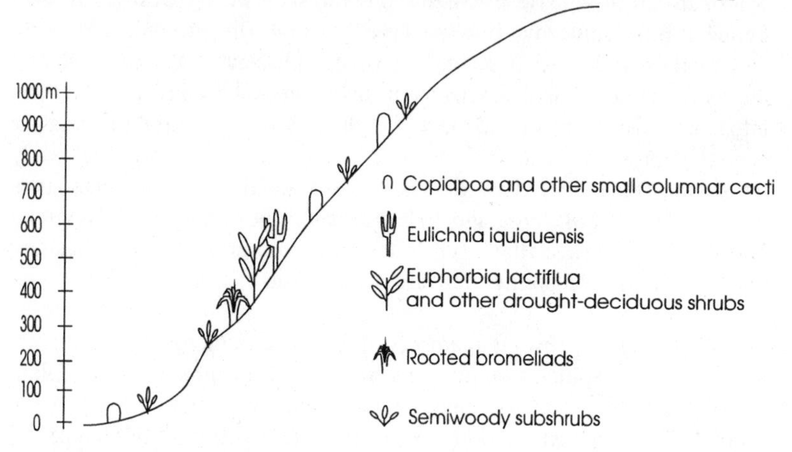

Figure 2 An idealized view of the lomas formations surrounding the town of Paposo, and their dominant vegetation.

mainly during the winter season. These hills are known as the lomas formations—true islands in a desert sea. Whether interception of fog moisture can sustain an ecosystem depends on many factors. An interesting dichotomy exists between the lomas communities of the northern Chilean and Peruvian deserts.

Chilean Lomas

The flora of the Chilean lomas is depauperate in comparison with that of the Peruvian lomas. This can be attributed to several factors, including the generally higher aridity of the Chilean coast and the more erratic supply of fog. In 1987, fog was so scarce that the epiphytic flora composed of tillandsias and lichens disappeared from the lomas near the mining town of Paposo. When there is fog, steep coastal hills, rising over 1500 meters near the town, manage to wring a substantial amount of fog moisture out of the stratus cloud layer before it moves into the interior valleys. This cloud layer lies at an altitude of between 300 and 800 m, which coincides with the vegetated band of the coastal hills. Figure 2 shows a diagrammatic representation of the coastal plain and the coastal hills. On the coastal plain, the only vegetation is broad stands of the small columnar cactus *Copiapoa haseltoniana*, and a few shrubs; the cacti and the shrubs barely cover 5% of the area. As we approach the fog zone at 300 m, we first notice large clumps of two species of rooted bromeliads—*Deuterocohnia chrysantha* and *Puya boliviensis*—among tall drought-deciduous shrubs. The best-watered area, centered at 550 meters, is dominated by the drought-deciduous spurge *Euphorbia lactiflua*, and the columnar cactus *Eulychnia iquiquensis*. These plants reach heights of 2 to 3 meters or more. There is a lush undergrowth of small shrubs and annuals (vegetation cover reaches 50% in this zone) attesting to the high relative humidity of this zone. In fact, during the winter and spring the humidity seldom falls below 80%, allowing epiphytic lichens and a species of tillandsia to grow among the branches of the drought-deciduous shrubs. As we approach the upper limit of fog at 900 m, the tall plants drop out. There the slope reverts to the control of *Copiapoa* and highly drought-tolerant shrubs and subshrubs that are members of the Nolanaceae and Asteraceae. As we climb above 900 m, vegetation becomes increasingly scarce, and at the altitudinal extreme of plant growth near 1060 m, only the small subshrub *Loasa fruticosa*, manages to survive. The Chilean fogs do not seem to deliver as much particulate matter as the Peruvian fogs; this may explain why most of the bromeliads that manage to make a stand have to be firmly rooted in the soil. Otherwise they might not be able to meet their mineral requirements.

As we travel north from Paposo on the way to Antofagasta, along the coastal hills and plains we encounter very little vegetation except for several species of *Copiapoa* and *Eulychnia iquiquensis*. Unlike the lomas around Paposo and Pan de Azúcar (which is to the south of Paposo and boasts over 120 species of plants), the immediate vicinity of Antofagasta supports only 60 species of vascular plants mostly concentrated along a dry river canyon known as Quebrada La Chimba. Just north of Antofagasta conditions become even harsher. One of the few signs of life there can be found on a prominent headland known as Cerro Moreno, which extends 1000 m above sea level. The vascular flora of the coast below the ridge is limited to scattered individuals of *Nolana peruviana.* As we approach the headland from the southwest along the coast we find an increasing number of *Nolana* species along with *Tetragonia angustifolia;* all are coastal halophytes that make do with salt spray for moisture. These halophytes increase in frequency along dry washes as we climb to the beginning of the fog zone at 600 m. The start of the fog zone is heralded by scattered clumps of *Copiapoa* and a depauperate assortment of shrubs such as the heliotrope *Heliotropium pycnophyllum*, a joint-fir, *Ephedra breana*, and a box thorn, *Lycium deserti.* As we approach the zone of heaviest fog deposition, between 800 and 1000 m, the cactus *Eulychnia iquiquensis* becomes the dominant vascular plant. The headland favors lichen and algal growth over that of vascular plants. Even *Eulychnia iquiquensis* does not grow vigorously here and is colored brick red by filamentous green algae of the genus *Trentepohlia* (order Trentepohliales) that grow profusely on its surface. The saxicolous and terricolous lichens of this headland have become a major part of the diet of a relic population of guanacos (*Lama guanicoe*) that have been stranded here by the drying climate since the wetter times of the Pleistocene. Along with the guanacos, a couple of nondesert plant species (*Acaena trifida* in the rose family and the spurge *Colliguaja odorifera*) migrated into this area during wetter Pleistocene times. They now exist as small populations disjunct from their nearest relatives in central Chile.

Why should the fog zone at Cerro Moreno be so propitious for terricolous and saxicolous lichen and algal growth, while the fog zone at Paposo is obviously better suited to epiphytic lichens and vascular plants? The answer may lie in the much lower frequency of condensing moisture at Cerro Moreno. Many lichens and some algae can extract moisture from nearly saturated air, whereas only a few vascular plants, such as some species of tillandsia, have been convincingly shown to accomplish this feat.

Copiapoa cinerea

Between Antofagasta (5-mm average yearly rainfall) and Arica (0.6-mm average yearly rainfall), the Pan American Highway turns inland into valleys that are blocked from receiving life-giving coastal fogs. These valleys are some of the most inhospitable deserts on Earth, and where plants manage to grow, their diversity is very limited. The only two life forms that seem to survive in such environments are the subshrub and the columnar cactus. Of the latter, the ones most studied for their amazing ability to produce small stands in areas that receive less than 5 mm of rainfall a year are several species of *Copiapoa*. It has been calculated that these cacti transpire water at a rate equivalent to 2.4 mm per square meter per year, and thus presumably survive where precipitation rates and soil properties conspire to produce this amount of infiltration per year. The difficulty lies in the random nature of this rainfall. Since periods of six years or more with no rainfall are not uncommon in this part of northern Chile, the calculated average rainfall is of no immediate benefit to a cactus caught in such a drought. To survive such periods a cactus must use its internal stores of water very frugally. For a mature *Copiapoa cinerea* 22 cm tall, the store might amount to 1 liter of water, and at the previous rate of transpiration this would last only 143 days; the store of a young plant 2 cm tall would last only 48 days. Since mature *Copiapoa cinerea* plants do exist in these areas, obviously they can control the duration and timing of stomatal closure better than almost any other plant. For the juvenile plants, however, the droughts in these areas might be fatal. It seems that the high probability of droughts that will completely deplete the water stores of seedlings and young plants is the most severe limiting factor preventing *Copiapoa* from colonizing more of this unpredictable land.

The *Copiapoa cinerea* cacti also possess other adaptations to the extreme desert they live in. A stand of these cacti have a propensity to point north at about a 52° angle, which is close to the angle for maximum quantum absorption for early spring along this coast. This adaptation might be geared toward accelerating growth (including flowering) in the early spring, when transpirational stresses are not as severe. By the time summer rolls around, the cactus can concentrate on limiting transpiration and water loss, having already achieved its reproductive imperative. This tilting adaptation is also seen in other columnar cacti in the Sonoran Desert, but in the Atacama the effect is so pronounced that it must have a highly beneficial result. Exposing only its apical meristem to the intense rays of the sun allows this cactus to shield its sides from excessive heating during the late summer and fall. Through this adaptation *Copiapoa* may be absorbing 20-30% less heat than randomly oriented cacti.

The Arid Divide

The arid divide I speak of is not caused by a sheer ridge, but it is the very lack of hills that produces an almost unsurmountable botanical barrier. The coastal stretch from Antofagasta in the south to Arica in the north has very little relief, and the cold moist air that comes inland encounters no obstacles that could intercept the cloud layer and harvest the fog. Since most of the ecosystems found on this desert coast count on, and in some cases absolutely depend on, the moisture extracted from fog, the lack of fog in this area is tantamount to having a desert within a desert.

The maximum rainfall recorded in the town of Arica was due to a single storm in January of 1918; it dropped a whopping 10 mm of rain that accounts for one third of all the precipitation received by that city over the period of record. Averaged over the years of record this is 0.6 cm a year, a truly meaningless number that serves to illustrate that rainfall in extreme deserts has two important components as far as life is concerned: amount and predictability. It is a pretty good rule of thumb that as the amount of rainfall decreases, its predictability also decreases. As we have seen, when precipitation becomes so unpredictable and meager, fog and other unconventional sources of moisture can become very important. But on this stretch from Antofagasta to Arica, even this source of life-giving water becomes fickle and unpredictable. From the mean cloudiness reported for the three cities of Antofagasta, Iquique, and Arica, it is expected that fog deposition in the fog zone above Iquique will be one twentieth of that in the fog zone above Antofagasta. Based on the same mean cloudiness data, the fog deposition in the fog zone above Arica will be one fifth of that around Antofagasta. The depauperate vascular vegetation at Cerro Moreno suggests that fog frequency north of Antofagasta is not sufficient to support significant fog-zone vegetation, even if the topography would allow fog interception. The coastal escarpment between Antofagasta and Arica only supports scattered individuals of several species of *Copiapoa* and *Eulychnia iquiquensis*. Near Iquique, in the Quebrada Huantajaya (a dry creek bed), only 12 species of vascular plants have been recorded. The only fog-zone community around Iquique is a population of *Tillandsia landbeckii* growing on pure sand at an altitude of 990 to 1100 m. Northward to Arica there is no vegetation to speak of. This stretch of land has had a pronounced effect on the biology of the whole desert region.

The character of the South American coastal desert changes as we travel from south to north. The species found in the south have affinities to those found in the areas of coastal Chile with a Mediterranean climate, while in the north the plants have Andean affinities. Of the 1200 species that populate the whole 3500-km coast, only 7% manage to cross the barrier created by the stretch of land between Antofagasta and Arica and appear in both the Chilean and Peruvian deserts. In fact, *Nolana lycioides* is the only species of *Nolana* that ranges over the whole coast from Antofagasta to Lima.

Peruvian Lomas

The Peruvian lomas are lush habitats that sometimes even support forests of introduced *Eucalyptus* and *Casuarina* trees. These trees increase the diversity of the community by serving as living interceptors of the fog droplets. In the lomas formations, the interception by the eucalyptus trees (amounting to 676 mm per year between 1944 and 1954, four times the amount of rain over the same period) allows plants of rain forest affinities to survive. For instance, epiphytic begonias, *Peperomia* (a member of the pepper family Piperaceae), and *Calceolaria* (a member of the figwort family Scrophulariaceae) abound on these lomas, and tillandsias have come to dominate the upper and lower limits of the fog zone. These tillandsias grow both on the trees and on the ground; but, as most tillandsias, they are not rooted and depend on the fog not only for water, but also for the nutrients that the particles delivered by the fog provide.

Other Epiphytic Habitats

Besides the true deserts we have just discussed, dry environments abound elsewhere, but they need not be labeled deserts on a map; in fact, such environments may be just meters away from the life-sustaining water they lack. *Tillandsia circinnata* grows on the buttonwoods and mangroves just above the intertidal zone on Sanibel Island, off the western coast of Florida. Some of these tillandsias accumulate 23% salt in or on their leaves (on a dry weight basis), yet seem to be perfectly happy in this location. These plants may be taking up moisture from the almost always saturated air in the mangrove forest.

In the vast mangrove swamps of the Indo-Pacific there are no bromeliads, but several other plant families have opted for the epiphytic existence. It is interesting to note that the families involved—the madder family (Rubiaceae), the milkweed family (Asclepiadaceae), and the polypody family (Polypodiaceae)—are also families with desert inhabitants. The only other family of Australian mangrove epiphytes, the orchids, are extremely dependent on their fungal symbionts in ways that remain unexplored, and since next to nothing is known about the fungi, I will refrain from discussing the orchids.

The mangrove epiphytes in the madder and milkweed families are all myrmecophytes, that is, ant-plants. These plants have established close relationships with certain species of ants. One of the better-studied associations is that of *Hydnophytum formicarium* with the ant *Iridomyrmex cordatus*. The plant produces two types of chambers, one smooth-walled and the other rough-walled. When the ant colonizes the plant it sets up housekeeping in the smooth chambers, and like any good ant, takes pains to keep this area clean. Many species of ants create refuse piles where they defecate and discard inedible bits including the bodies of fallen comrades. *Iridomyrmex* does the same thing; it just does it in the conveniently provided receptacle of the rough-walled chamber. The increased area of this chamber provided by the rough walls possibly aids the plant in absorbing nutrients and water faster, just as the villi in a vertebrate's small intestine do. What is known for certain is that radioactively labeled nutrients defecated by the ants do make it up the stem of the plants, and the plants with *Iridomyrmex* colonies produce more offspring than plants with no colonies or colonies of a different species of ant. Myrmecophytes seem to have evolved mainly in epiphytic situations in which water and nutrients are scarce, as they are in open canopy forests such as mangals.

Even though the epiphytic habit is undeniably stressful in terms of plant-water relations, the stress is mainly due to the lack of a moisture-retaining substrate, not to a lack of precipitation. In true deserts, however, lack of precipitation is the overall controlling factor, with other stressors giving each desert its unique character.

The staghorn fern (*Platycerium bifurcatum*), one of the largest epiphytic ferns, can grow to more than one meter across.

3 Your House is My House

Introduction

This chapter deals with commensals—organisms living with, on, or in another organism. In general, a commensal organism can benefit, hinder, or play no significant role in the survival of its host. Some of these commensals have embarked on a difficult road; they have become guests in the exposed nests of birds, or invaded the dry environments of human habitations. These environments are every bit as dry as deserts, and we will see that obtaining water becomes a major driving force in the evolution of commensals and external parasites.

Some commensals have become so closely tied to their host that they cannot survive without this close association. Of this subset, the ones that derive nutrition directly from the host without providing any benefit to it are normally known as parasites. For parasites, be they external like fleas and lice, or internal like tapeworms, the severe limits (conditions) imposed by the host are the powerful driving forces behind the evolution of these unwanted guests. When we study parasites we see evolution moving at a rapid pace, and in this chapter we will encounter several creatures that are undeniably moving towards a parasitic existence, but do not seem to have made the final transition.

For internal parasites, water is no longer a problem, but in the digestive tract of large organisms like the horse, oxygen many times is in short supply. Some of these parasites behave as transients, and their adaptations to low oxygen are of a stop-gap nature, while others have truly become masters of the anaerobic realm.

It might seem strange to follow a discussion of deserts with one on commensalism, but it turns out that much of what we know about the physiological adaptations of insects to lack of water comes from studies of commensal insects. For example, in the Namib, as well as other deserts with extensive sand seas, some of the most plentiful animals are bristletails. Unfortunately, almost nothing is known about these insects. On the other hand, we know a lot about the habits and physiology of your friend and mine—the common silverfish.

Commensalism is not restricted to animals. Algae not only may be found growing on other plants (as mentioned in the previous chapter) but may be found growing on a variety of animals as well—if the moisture and light conditions are adequate. In this case they are referred to as epizoic.

Many marine and freshwater animals are coated with a variety of such algae. An even closer plant-animal association is found in some aquatic invertebrates, where the animal serves as host to an endozoic (internal) alga.

Few algae could be called parasites, since almost invariably, the epizoic types are innocuous to the animal hosts, and the endozoic types supply their hosts with food and help in $CaCO_3$ deposition. On the other hand, epizoic animals have taken the first step on the road to becoming ectoparasites.

The type of commensalism where an organism lives within the nest of another is known as inquilinism. Even though some biologists prefer to narrowly define inquilines as those organisms that are commensals in the nests of social insects, I will keep the broader definition. When the relationship is mutually beneficial it may be termed a symbiosis; if it is detrimental to one of the organisms, then it is parasitism. Sometimes, the relationship is hard to classify because of our lack of knowledge of the costs and benefits to each member, and sometimes because it just does not fit into our scheme. We will encounter examples of both kinds of association in this chapter. We will also find commensals in various stages on the road to true parasitism, and we will concentrate on those aspects of the host that make the commensal's (or parasite's) environment difficult ground to conquer.

Algae as Commensals and Symbionts

The most famous case of a plant/animal symbiosis is probably that of the zooxanthellae (intracellular algae) of hard corals. These algae grow nowhere else but in the living tissues of several families of hard corals, and their only method of reproduction seems to be by hitching a ride inside the sex cells of the corals. When these corals spawn (they do this en masse all over a particular reef), each egg carries inside it an algal cell that will divide at the same rate as its embryonic host. Other algal-invertebrate symbioses include those of zooxanthellae with some invertebrates such as sea anemones, the flatworm *Convoluta*, the polychaete *Eunice gigantea*, and the giant clam *Tridacna*. In the fresh-water realm, the hydra can internalize a green alga (zoochlorellae). The green alga-hydra combination is then known as *Chlorohydra viridissima*. Because the eggs of this strain of hydra carry the algae, the symbiont is inherited.

The nudibranch of the genus *Placobranchus* has developed a different kind of association with an alga, although in this case the algal partner is not a whole organism but just part of one. This slug feeds on the common green alga *Codium*, by piercing individual cell walls. Instead of digesting the whole meal, there are special cells present in the slug's alimentary tract that "save" the chloroplasts of the digested cells. These chloroplasts are able to function perfectly well inside the slug's body, and the animal has a significant amount of control over them. The slug is able to relocate the chloroplasts to its dorsal surface, so that they can be in the best spot for photosynthesis. Unfortunately, chloroplasts are not cells (although they evolved from free-living organisms), and they are unable to multiply without external cellular machinery that only their original host alga can provide. As the chloroplasts within a slug's cells age and die, no new ones take their place unless the slug ingests more algae and starts the cycle anew.

Animal Inquilines

Many vertebrates build nests to protect themselves and their young from the elements. We in particular like to build protective enclosures for ourselves and our families, and for our possessions. Some organisms have moved into these ready-made shelters unannounced and many times unwelcomed. Here they have found food as well as protection from predators. Some of these organisms become dependent on this association, feeding on the food or excrement of their hosts—these organisms are called inquilines. In the case of humans, inquilines find a varied array of substances on which to feed, including woolen garments, wood and paper products, dried foods, and even the remains of the interred. We, unlike many other creatures, are quite perturbed by unwanted guests. Of course, to us they are pests, although only a few cause real damage to property.

Inquilines have discovered an almost unlimited food supply. The reliability of this supply has been the driving force for the development of sedentary lifestyles in many of these organisms. Some inquilines face the same evolutionary choices that small desert animals face: move between the source of water and the food, or stay with the food and be creative about obtaining water. Some inquilines extract all the water they require from the food they eat; but if the food is particularly dry, such as stored grain, bits of feather, or the wool of your prized afghan, then these animals must come up with mechanisms to get and conserve water. As we will see later, some invertebrates such as the silverfish (*Lepisma* spp.) and the cigarette beetle (*Lasioderma serricorne*) have independently evolved complicated organs to extract water out of the very air!

Feathers are a rich source of protein, but it is a slippery slope from feeding on molted feathers to feeding on them while still attached to their owners. Once the association between inquiline and host becomes so close that the inquiline spends most of its time on the host, the boundary between inquilinism and parasitism is blurred. Even if the guest does not actually feed on the host's living tissues, it may debilitate the host by causing irritation, mechanical damage, and obstructions that may be life threatening. Once the guest starts feeding on the host's living tissues, there is no doubt that it has become a parasite.

Internal inquilines and parasites have found a plentiful and reliable supply of food and moisture. This cornucopia has been their evolutionary incentive to find novel solutions to the lack of oxygen often encountered inside their host. Intestinal parasites are protected from external fluctuations in temperature and from desiccation, but utopia, the intestine is not. Helminth parasites must put up with extremely low oxygen levels. However, these worms have managed to evolve a set of biochemical pathways that allows them to use carbon dioxide to get rid of excess electrons (one of the jobs oxygen used to do for their ancestors); they assimilate carbon dioxide just as plants do.

Inquilines and Man

Every human habitation has several different species of uninvited guests. These make a simple food web that depends on the habits of the host. In my house silverfish are the threat, together with the odd earwig that ventures in from outside when rains have flooded its home, or when it seeks a moist site on a dry winter's day. Every few years there might be an invasion of millipedes when the winter is particularly mild and moist, but these are doomed to die of desiccation, as are most trespassers into energy-efficient homes. Other animals have very restricted ranges, and their whole world may be contained in the soil of a flowerpot. This is the case for the springtails that are sometimes seen hopping about in flowerpots that are kept overly moist; however, springtails are not well prepared for an existence inside a modern human habitation. But if one inspects a flowerpot closely, the soil will reveal mites and sometimes even book lice. These animals are feeding on the fungi found in the soil or on the surface of the plants, not causing any harm and unable to spread from their moist island to any other, except on the spout of a watering can.

Where there is prey there will be predators. In temperate countries the most common predators in houses are spiders, mostly jumping and wolf spiders; they are at the top of the house-inquiline food web. It is no coincidence that we will find a similar food web on high Himalayan and Andean slopes, for in all these places the controlling factors are similar.

The Silverfish (Thysanura)

The limiting factors that keep the house inquilines already mentioned from ranging farther afield are lack of food and moisture. Silverfish are perfectly happy chewing on the papers in my office or the protein-rich glue of the bindings in my books, and for them my office must be a silverfish's Sybaris. For the earwig, venturing into a house is a dangerous thing unless there is a moist sanctuary that the animal can go to periodically to replenish its moisture. In houses with no leaky pipes or moist basements, water is the most precious commodity for any inquiline. Silverfish are not as constrained by water sources as other inquilines, and they can be found alive even weeks after falling into a cup or a pencil holder. If left alone, a trapped silverfish will undergo a series of retrogressive molts during which it becomes smaller at each passing molt. If food and water are not forthcoming, after a while the animal's alimentary tract will start to atrophy; eventually, even if it can escape its prison, it will be unable to feed and will die. If an earwig found itself in the same predicament, it would quickly dehydrate and die. What makes silverfish so much better able to withstand water deprivation than earwigs? The silverfish and its now-rare relative the firebrat are adapted to survive dry but constant conditions. They are the lucky owners of one of the few organs in all of animated creation designed for the absorption of water vapor from air. Their version of this anatomical and physiological marvel is also one of the most advanced designs we will encounter.

The organ used by the firebrat (*Thermobia domestica*) and the silverfish is capable of absorbing water vapor from air to a relative humidity level of about 45% (depending on the temperature of the environment). The firebrat's temperature requirements used to make it a common pest in bakeries, boiler rooms, and around often-used stoves and fireplaces. Now that fireplaces are less common and bakeries are much more sanitary, the firebrat has become relatively rare. This creature requires temperatures of 42°C to complete its development, hinting at an ancestral hot desert home. In fact, thysanurans are common in the sandy deserts of the Middle East. They have hardly been studied in these habitats, even though they probably are very important components of the biota of these deserts.

The firebrat possesses one of the most effective water-vapor-absorption organs known, a very useful organ for a creature living at such high temperatures. The organ itself consists of three sacs that contain a granular material and are evaginated from the anal wall. Several theories have been proposed to explain the mechanism of water-vapor uptake. What follows is my own version of how the mechanism might operate. First, the insect ventilates its anus to bring in moist air from outside. Then, the water vapor in this air is absorbed by the granular material in the anal sacs. In the third step, this granular material must be made to release the bound water; this is accomplished by secreting ions into the sacs. This action collapses the swollen granules and releases the bound water. In the last step, the epithelial cells that released the ions reabsorb them, and the water is dragged along (a process known as electroosmosis and often used by soil engineers to dry out sediments), and it is also reabsorbed.

Stored-Product Pests

Beetles

The beetles that attack stored products can be separated into two classes based on their preferred foods. The darkling beetles (tenebrionids) prefer grain and dried plant materials, whereas the carpet and larder beetles (dermestids) prefer dried animal products (like horn, bone, pelts and skins) and dried meats. The death-watch beetles and their relatives (anobiids) are somewhere in between in terms of food preferences.

Darkling beetles have some remarkable forms that do not become pests, such as the rugged darkling beetles sometimes found in Mexican curio shops. These beetles can survive for months on a shelf or in a jar, indignantly festooned with small bits of colored cloth or rhinestones, until a tourist buys them; hopefully, they can then escape their indignity. We met other darkling beetles in the Namib Desert, where their unique behavioral adaptations let them harvest water out of passing fogs. The rogue members of the family include the yellow mealworm (*Tenebrio molitor*) and several other smaller flour beetles (especially those in the genus *Tribolium*).

Dermestids prefer sources of animal protein but will not turn away from materials of other sorts. Their larvae are the bane of museum collections and can completely destroy insect collections. They will also disfigure and otherwise ruin anything made of fur, feathers, or hair. In contrast to the delinquent larvae, the adults placidly feed on flowers, taking pollen and nectar. Some dermestids are being made to repay the damage inflicted by their kin in museums throughout the ages, since they are now enlisted by museums for the job of cleaning carcasses so that the bones can be mounted for display.

Death-watch beetles feed almost exclusively on plant material, with many making seasoned wood their home. It is this habit that has branded them "death-watch beetles." In the distant past it was common to hear a ticking sound coming from the woodwork of deathbeds, a sound believed by some to be an omen of death (or so the story goes). This ticking was caused by the larvae of the beetles trying to communicate their whereabouts to other larvae inhabiting the log, lest their tunnels run together (unbeknown to them, the other larvae were by then probably in the kitchen table).

Some death-watch beetles are extremely long-lived for insects, and stories abound of larva that, having taken up residence in a log that later became someone's desk, kept the new owner company with a comforting tap, tap, tap for many years. Their very long life (over 12 years in some cases) is a result of their very poor diet, since they are unable to digest cellulose and must process large quantities of wood to extract enough nourishment to grow even at a modest rate.

The two most destructive members of the Anobiidae are the drugstore beetle (*Stegobium paniceum*) and the cigarette beetle. Unlike most other members of the family, these two species have extremely catholic tastes and have, at one time or another, become pests of tobacco, dried plant materials of all sorts (from raisins to books), drugs, black and red pepper, pyrethrum powder, and even strychnine.

In the Anobiidae, the water-vapor-absorption system is completely different from that of the firebrat. It involves a tube that starts at the Malpighian tubules and runs along the thin intestine and the rectum to finally open next to the anus. Of the Anobiidae studied to date, the one with the best water-vapor-absorption system is the cigarette beetle larva. This larva can extract water vapor from air with a relative humidity of 43%. This trait is also found in the larvae of darkling beetles, but in them it is not nearly as efficient. Strangely, only beetle larvae are able to accomplish this feat, and no adult has been found capable of even weak water-vapor absorption.

Moths

As is done with beetles, pest moths can be classified according to the stored products they infest, and again, moths also prefer either animal or plant material. Clothes moths and their relatives prefer the keratin found in wool or fur, while meal moths concentrate on stored grain or meal. Even though their food contains only from 1 to 5% water none of these moths have developed organs for water vapor absorption and rely mostly on behavioral adaptations, seeking dark and humid closets or chests to keep from desiccating. The clothing on which they live acts as a humidity buffer, absorbing water on humid days and slowly releasing it during dry days. In this way they are protected from extreme humidity fluctuations.

The Book Lice (family Psocidae, order Psocoptera)

Book lice are diminutive insects that spend their lives on the bark of trees or feeding on the fungi that grow on old books. Sometimes they find their way into stored flour, and since their color matches that of flour, they are only detected when tiny specks of flour start to run away. These insects are in many respects similar to lice, but on a much reduced scale. Some are commensals in birds' nests or in burrows of mammals, and a few have been found on the owners of these habitations.

From Inquiline to Parasite

A few different lineages of insects have become true parasites of vertebrates, but almost as many are still in transition, a journey that will make them completely dependent on their host and the environment the host provides for them. In some cases this environment is fairly benign; in others it is harsh, for though food is in plentiful supply, other necessities, such as water and oxygen, may be quite scarce. In the latter case, the host provides an environment that spurs rapid evolution of the organism that is able to conquer the scarcity. If it is able to adapt, such an organism will be rewarded with an inexhaustible food supply.

The Earwigs and Their Relatives (Dermaptera)

Based on their usual role as harmless trespassers, it is difficult to think of earwigs as predators or as parasites, but this is exactly what some of the species have become. Their presence in the aeolian zone of the Andes (see chapter 10) suggests that they are able to feed on dead and dying insects, the major source of nutrition available in this windswept habitat. This habit has been taken to an extreme by those species that share caves with bats. Caves that have been inhabited for long periods of time can have large populations of insects in the guano at the base of the bat roosts. Sometimes the whole mass writhes and undulates like a single living organism. Most of the insects composing this mass are feeding on the guano itself, but others specialize on any fallen baby bats or on the already mummified corpses of adult bats. A bat's upside-down resting posture has stimulated the evolution of a normally closed grasping mechanism in its feet. Because of this, a relaxed bat can hang on without any exertion, and by the same token, a dead bat may remain attached to the wall of a cave until it is bumped off or the ligaments holding its claws in a closed position deteriorate. It is a simple path from feeding on fallen bats, to feeding on dead bats while they still hang from the roost, to feeding on neighboring moribund bats. This is the path some earwigs seem to have taken on the way to becoming external parasites of bats.

Arixenia is an almost typical earwig. It is 2 cm long; it has a pair of small compound eyes, and it has well-sclerotized cerci that look like the cerci of smaller, more delicate earwigs. This genus contains only two species: *Arixenia jacobsoni* and *Arixenia esau*; each species was first found in a different cave in Southeast Asia. Whereas *A. jacobsoni* ranges from the Philippines to Java, *A. esau* has been found in only one other place—inside a hollow tree where a colony of naked bats (*Cheiromeles torquatus*) had made a home.

Arixenia jacobsoni was discovered by Jacobson in a bat cave in southern Java. That this species' discoverer did not hold the insect in high esteem can be gleaned from the following account as recounted by Burr and Jordan:

> The most conspicuous insects inhabiting the cavern are ... the Earwigs ... they crawl in countless numbers on the surface of the guano and everywhere on the rocky walls. Evidently they live on the various larvae feeding on the guano, but besides this they are constantly waging a terrible war against each other, the victors devouring the bodies of their slain mates ... A more loathsome spectacle than these thousands of ugly, hairy creatures, running about hither and thither, fighting and devouring each other, can hardly be imagined. (Burr and Jordan, 1912)

It is an ironic bit of entomological justice that the name of this creature is *Arixenia jacobsoni* in honor of its discoverer.

Arixenia esau was initialy discovered in a cave near the Niah River in northwestern Borneo, in a colony of naked bats (*Cheiromeles torquatus*) by the British zoologist Lord Medway. These insects were not present on live animals captured away from the roost, only on bats found on the floor of the cave. These bats, if alive, were invariably sick and dying, and the earwigs seemed to feed on their skin. Lord Medway tested this hypothesis by picking up several of the insects. They immediately started feeding by scraping the skin between his fingers and moving upwards as they fed. Other *A. esau* were found preying on their compatriots in the dung: adult darkling beetles. They also fed on dead insects; this made them seem like opportunists more than like parasites. As mentioned earlier, the only other place where *A. esau* has been found is inside a hollow tree—the home of a colony of naked bats. In this instance the earwigs were collected from the bodies of bats in the roost; this strongly suggested that *Arixenia esau* is a facultative parasite of those naked bats. *Arixenia jacobsoni* was also collected from that tree, but the few individuals that were found occurred only in the guano.

Heminirus is another earwig relative that provides clues about the path insects have taken on their way to becoming parasites. The nine species of this earwig-like genus are parasites of pouched rats in tropical Africa. What makes this insect unique is its apparent taxonomic isolation, since it does not appear to be closely related to either earwigs or cockroaches but looks a little bit like both. It lacks wings and eyes, but it has, as earwigs do, two cerci at the end of the abdomen. *Heminirus* is viviparous (as is *Arixenia*), and this insect has even developed a placenta-like structure to feed its unborn young. It has radiated, along with its host, into arid regions where life is harsher for the host, and by inference, for the parasite as well. The parasites found in these arid regions has a curious anatomical detail: it is able to close off its anus by using the last sternite to lock onto the ventral surface of the last tergite. This modification is very likely an adaptation for restricting water loss. But, as we have seen already, the anus is the main site of water-vapor absorption in several orders of insects. Could this be the case with *Heminirus* also? Its diet of thin slices of its host's epidermis together with the spores and sporangia of a skin fungus suggests that these insects—as the biting lice we will shortly meet—need a supplementary source of water.

Other Insects in Transition

If we now travel to the American tropics we find another transitional situation in which a host has become a safe and nutritive haven for an insect. I am referring to the relationship of the three-toed sloth (*Bradypus* spp.) and the two species of moths in the genus *Bradypodicola*. As mentioned earlier, the adults and larvae of these moths live among the thick hairs of the three-toed sloth. The adult moths, having the standard lepidopteran drinking-straw mouth, must feed on skin secretions (some nonparasitic tropical moths do the same). The larvae seem to feed on the hairs themselves (as if they were clothes moths feeding on a fur coat) and on the algae within the pits. If we want to find those insects that have made the final transition from feeding on the dead to feeding on the living, we must look to the flies, fleas, and lice.

Flies belonging to several different families have become external parasites of vertebrates. Of these, only one is of interest to us, because it shows the path some insects may have taken on their way to becoming external parasites. This fly is *Carnus hemapterus* of the family Milichiidae. This fly's larvae feed on the rich organic materials found in birds' nests (as is the case with many other fly species). The adults, on the other hand, lead a parasitic lifestyle among the feathers of the nestlings. As soon as this fly takes up residence on a bird, it loses its wings and its abdomen becomes distended. However, its mouthparts are not yet well suited for piercing the skin. It is a good example of an insect that has recently embarked on a parasitic lifestyle and may in time become as morphologically exceptional as the pupipara (the tsetse fly and its relatives).

The Journey Inside: Myiasis

Myiasis is the infestation of a living vertebrate's body by flies (Order Diptera). There are many different forms of myiasis; the most serious ones are produced by the scourges of domesticated stock—warble flies and screwworm flies. Of lesser economic importance but more interesting from our standpoint is the myiasis caused by the ingestion of fly larvae. Our intent is to focus on those forms of myiasis that provide narrow examples of the challenges faced by an organism that takes up residence inside another.

We will start with pseudomyiasis. In this form of myiasis, the maggot involved is simply going about its business, with no expectation of entering the body cavity of a vertebrate, and once there it is hard put to get out of that labyrinth. Such is the case of the cheese-skipper, the larva of the fly *Piophila casei*, which sometimes still infests meats and cheeses. This maggot is sometimes accidentally ingested with the food it infests, and it can cause extreme retching as it writhes and thrashes in the victim's digestive tract, causing damage with its mouth hooks. If the victim's stomach acid is weak because of illness or age, the maggot can survive transit through the stomach and eventually emerge at the end of the digestive tract (if the victim has not vomited it first). This type of infestation is known as pseudomyiasis because it is not the intention of the larva to be swallowed, and once in the digestive tract, it does not undergo any development. It strenuously objects to the low oxygen and high carbon dioxide concentrations, and it tries with every movement at its disposal to get out.

A different form of accidental infestation can be termed facultative myiasis. In this case the fly does not customarily infest vertebrate animals, but if it happens to be swallowed by a vertebrate, it makes the best of the situation and may even set up temporary residence in the intestine. An interesting case that was popularized in Oldroyd's book *The Natural History of Flies* recounts the story of a man living in Burma that accidentally ingested larvae of *Megaselia scalaris*, a cosmopolitan humpbacked fly (or as Oldroyd liked to call it—a coffin fly) that feeds on many things including emulsion paints. These larvae found adequate accommodations in the man's intestines and took up residence there for over a year. Over this time, larvae of different ages, pupae, and even adults were voided by this chagrined patient. Some have argued that the flies were mating in the man's intestines; others have labeled that an impossibility. Be that as it may, the survival of a population of nonparasitic flies in a man's intestines for over a year is a truly astounding physiological achievement for this fly—albeit a gruesome one.

Oldroyd recounts observations conducted by a colleague of his on the life cycle of another cosmopolitan humpbacked fly, *Conicera tibialis*. These flies were observed coming and going from the carcass of a dog buried about a meter underground. The adult flies excavated their way to the surface of the ground where they mated, and without any hesitation the females started digging down toward the corpse, were they laid their eggs. This and other phorids are found on and in corpses that have been interred for a while, and it is unlikely that the female phorid is able to lay eggs on the corpses before burial. A biologically accurate rendition of Edgar Allan Poe's poem "The Conqueror Worm" would be entitled "The Conqueror Maggot"!

Stomach Bots

Stomach bots are a set of bot fly larvae that are inquilines of ungulates and elephants. Although technically not parasites, they do enough harm to the host to be considered as such. The adaptations of internal parasites to lack of oxygen have been independently evolved to suit the individual species' life cycle. Some of the most ingenious are the adaptations of the stomach bot flies (Gasterophilidae). These flies attack large ungulates like horses, zebras, and asses. They also attack rhinoceroses and elephants. They do not attack ruminant ungulates like cows and sheep. Ruminants may be immune to these flies because the act of chewing the cud would be fatal to the fly larvae (the bots). Besides, the anaerobic conditions in a ruminant's gut are much more pronounced that in other ungulates. These factors may have hampered the evolution of a stomach bot specializing in ruminants.

The bot flies of the genus *Gasterophilus* parasitize the Equidae (the horse and its relatives); the horse alone has four separate species attacking it. The different species differ in several aspects of their life cycle, and they do not compete directly for resources. They also have different modes of infecting the host. For example, *G. precorum* females lay thousands of eggs on vegetation. These eggs can lie dormant for several months until a horse feeds on that vegetation, at which time the eggs hatch and the larvae burrow into their host's soft palate; later they move to the stomach to complete their last larval stage. The females of *G. haemorroidalis* oviposit on the horse's lips. As soon as the eggs are wetted by saliva, the larvae hatch and burrow into the lips; later they make their way to the stomach and duodenum, and before pupating, they spend their last larval stage in their host's rectum (hence their name). The *G. intestinalis* female finds a horse and forcefully attacks the front quarters of the animal making repeated bombing raids. At each successful bombing raid the fly leaves one egg stuck to the horse's hairs. The eggs are set to hatch the minute they are rubbed by the horse's tongue. The newly emerged larvae burrow into their host's tongue and later take up residence in its stomach or duodenum. In the case of *G. nigricornis*, the female lays its eggs on the horse's cheeks, where they hatch spontaneously without any help from the host. The larvae then move to the corners of the mouth, burrow into the epidermis, and make their way to the inside of the cheeks.

These four species may seem to be directly competing with one another inside their host, but they manage to reduce competition by simply occupying different areas of the horse's mouth. Once in the stomach, the larvae of all four species attach themselves to the stomach lining with powerful hooks. The larvae are actually internal commensals and feed on the food that passes by; they do not eat the horse's tissues, but can debilitate the horse if their population is so large that they consume a substantial portion of the ingested food. A more serious threat in this case is the gastric bleeding caused by the many hooks embedded in the host's gastric lining, and the possibility of intestinal blockage when the larvae finish their feeding phase and release their hold on the lining. Once the larvae release their hold, they pass out the anus and pupate just below the ground surface.

The stomach bots that infest horses have been intensively studied. This research has shown that the larvae of these flies have evolved several elegant adaptations to their anaerobic odyssey. For starters, these larvae contain insect hemoglobin that allows them to utilize oxygen at the very low partial pressures sometimes found in the stomach of a large animal (they can survive in the complete absence of oxygen for up to 17 days). While feeding in the alimentary tract of its host the larvae deposit large energy reserves. If they were to spend their whole life in the stomach of a horse, then carrying around this reserve would be an unnecessary burden, but they will eventually pass through the gut and emerge at the other end as pupae. In this case the larvae's energy store becomes the adults' perfect fuel, which allows bot flies to skip adult feeding altogether.

The bot flies that attack rhinoceroses belong to the genus *Gyrostigma*. They are large creatures both as larvae and as adults. Their habits seem to differ little from those of *Gasterophilus*, except that the infestations normally number in the hundreds of individuals. Oldroyd considers this a likely reason for the rhinoceroses' reputation of being chronically ill-tempered.

There are several other bots that are inquilines of rhinoceroses and of elephants; they have been studied little, except for the elephant bot *Platicobboldia loxodontis*. The female of this fly lays its eggs at the base of the tusks of African elephants, and the newly hatched larvae enter the elephant's mouth. Instead of hooking into their host's stomach lining, these larvae freely roam the inside of the stomach, and when they are ready to pupate, they simply go back to whence they came, congregating on the underside of the tongue. The elephant then coughs the whole writhing ball of maggots onto the ground where they disperse to find sheltered crannies in which to pupate.

Fleas

Fleas are less wedded to their host than lice are, and many fleas do not really care where their next meal comes from. Some bird fleas may feed on over one hundred different kinds of birds, with a few mammals thrown in for good measure. Adult fleas' catholic tastes may be due to both their active locomotion and jumping and their ability to survive away from any host for months at a time, especially if the temperature is just above freezing. However, sometimes fleas, like lice, settle down. Then they are at the mercy of the evolutionary pressures produced by one host. This is the case for two fleas that infest the American cottontail rabbit. One of the fleas, *Cediopsylla simplex*, is found only on the rabbit's head, while the other, *Odontopsyllus multispinosus*, is found on its back and flanks. A few fleas, called stick-tight fleas, have adopted habits even more sedentary than those of most lice. Two of the most common stick-tight fleas are the poultry flea (*Echidnophaga gallinacea*) and the rabbit flea (*Spilopsyllus cuniculi*). The female stick-tight flea permanently attaches to its host's head by embedding its mouthparts in the host's flesh. The flea's mouthparts soon enlarge and its legs atrophy, cementing the bond between parasite and host. The rabbit flea is also unusual in that its reproductive cycle is completely tied to that of the host. Until a female rabbit flea feeds on a pregnant host, its ovaries do not mature; the trigger seems to be the elevated levels of corticosteroids and estrogens in the host's blood. Just after the host's parturition, the fleas move to the newborns, and this is the only time they mate. Mating seems to be triggered by the elevated levels of corticosteroids and growth hormone in the newborns' blood. The female fleas then start their egg-laying period which lasts about ten days. After this, they return to the lactating doe, and her changed hormonal state probably causes the fleas' ovaries to regress.

Humans are not spared the attentions of stick-tight fleas, and a South American flea, the chigoe (*Tunga penetrans*), has made itself a pantropical nuisance. The female burrows between the toes, or under the toenails, of its host, where it stays for the rest of its life. Because of its small size (less than 1 mm long), it is seldom detected as it makes its way into its newly found home, but once on easy street, its abdomen distends and becomes a plump 5 mm in diameter. The only part of the flea that protrudes above the surface of the host's skin is the genital opening, waiting for any male to brush by. Obviously *Tunga* must dispense with any sort of courtship behavior, but this is the usual state of affairs for fleas. Species-isolating mechanisms in fleas are not behavioral; they are structural. In each flea species, the male and the female genitalia are like a lock and key. This morphological approach is not as effective as a behavioral approach, and hybrids have been recorded.

Once the female flea lays its eggs, which it might do loosely on the host, or in the host's nest, there is a certain amount of parental care—of sorts. Just before mating, the female rabbit flea starts feeding to excess and defecating a bloody pellet every few minutes. These fecal pellets drop off the host and dry on the floor of the nest—a ready source of food for the newly hatched larvae. Many, if not most, flea larvae depend on this concentrated pelletized food source, and some of the more motile fleas have taken this parental care even further. The adult of one of the many rat fleas, *Nosopsyllus fasciatus*, spends a substantial part of its time roaming the host's nest, where its larvae hide. If an adult flea passes close to a larva, the larva clamps onto the pygidial region of the adult's abdomen. This clamping action stimulates the adult to defecate, and the larva feeds as the slightly used blood passes out of the adult's anus. These larvae have food and water delivered to them; however, most flea larvae ingest very dry material and must find a different source of water.

The environment of the larvae is the most important factor controlling the distribution of flea species, but probably the most susceptible stage of a flea's life cycle is the pupa. Like a contortionist in a straight jacket, to emerge from its puparium, an adult flea must make certain movements. At this time the flea is most susceptible to desiccation, since even a slight shrinkage of its body will make its movements ineffectual.

Depending on the host's species, the nest of a vertebrate host can be a very demanding environment. The nests of mammals are relatively congenial places. They are often occupied year-round by their owners, and they are often located underground, which makes them relatively cool and moist. The nests of birds are occupied for a shorter time by their owners and are usually more exposed than those of mammals. This may be the main reason fleas are more successful as parasites of mammals than of birds, and even just among birds, those that nest in more exposed conditions have fewer fleas. For instance, rockhopper penguins nesting in sheltered spots on Macquarie Island in the southern Indian Ocean are parasitized by *Parapsyllus magellanicus*, whereas those that nest in exposed situations have no fleas. The main reason for this is probably temperature, since few flea eggs and larvae can survive temperatures lower than -5°C. In other instances the controlling factor must be humidity. For example, one survey of birds' nests built among the branches of trees failed to find a single flea, probably because of the low humidity encountered in such nests.

A flea-infested house that is left empty for a while is a CO_2-activated bomb waiting to explode. The flea larvae that were left behind pupated, and are now diapausing adults waiting for the kiss of carbon dioxide to awake them from their sleep. They will emerge en masse at the slightest breath of a dog, cat, or human. The only requirement is a puff of warm CO_2-laden air to tell the waiting adult fleas that a meal is at hand.

In the case of birds' fleas, their emergence from their cocoons is triggered by the warm days of spring. For a few days after emergence they shun light and mate in the dark recesses of the old nest. After this period they reverse their behavior and seek a high, well-lit place. This behavior leads them to perch on the tops of vegetation surrounding the old nest. Once there, any sudden drop in light intensity will cause the fleas to jump, in hopes of catching a ride on a bird going back to its new nest.

Lice

Unlike fleas, which though directly dependent on their host as adults, lead a relatively free juvenile existence, lice are strictly parasitic and spend every stage of their life cycle on their host. For the most part, lice are host specific; each species is not able to grow and reproduce except on one mammal or bird. Some species take this specificity to an extreme, and they take up residence on only one particular area of the host, from which they seldom wander. Each species of lice is so closely tied to its host species that if a louse finds itself on a different host it might as well be on another planet. In such a foreign environment it will have difficulty even walking—its tarsal claws having evolved to grasp a different sort of hair—to say nothing of attaching eggs to this alien substrate. Consequently, lice are the quintessential "islanders"; separation from their host is invariably fatal to them.

There are two types of lice: the biting lice (order Mallophaga) and the sucking lice (order Anoplura). Some taxonomists place both types of lice in the order Phthiraptera, and then Mallophaga and Anoplura are relegated to the status of suborders. The debate as to their kinship revolves around their mouthparts. Biting lice make their living by chewing (maybe it would be more accurate to call them "chewing lice") on their host's feathers, or scraping the top layers of its epidermis.

We will be mainly concerned with the biting lice, for even though some of these lice compensate for the extreme dryness of their habitat by having a blood meal whenever they can, others find more unusual sources of moisture. Probably most mallophagans do take advantage of any wounds on the surface of their host's skin to ingest blood or exudates; a few even pierce the soft pith of developing quills of birds in order to extract blood. It has been reported that one genus pierces the skin of birds with its mandibles in order to drink blood. Almost all biting lice have run-of-the-mill chewing mouthparts; the one exception is *Trochiloecetes*, which has evolved piercing mouthparts, though they are different from the piercing mouthparts of the Anoplura. There are also a few biting lice that have made a move toward an endoparasitic existence by taking up residence inside the pouches of pelicans, where they must feed on mucus and blood, since feathers are unavailable. But even these lice are compelled by their surface heritage to return to feathers in order to lay their eggs.

The other group of lice, the sucking lice, derive all of their nourishment and water from blood meals they extract from their host, and therefore, they do not suffer the extremely dehydrating conditions encountered by the biting lice. While biting lice parasitize both birds and mammals, sucking lice parasitize only mammals. Strangely, the most primitive mammals, the monotremes (the duck-billed platypus and the several species of echidna) are not parasitized by either group of lice. This is also the case for the mammalian orders Chiroptera, Pholidota, Cetacea, and Sirenia. Two of these, the Cetacea (whales and dolphins) and the Sirenia (dugongs and manatees) are totally aquatic; their aquatic environment may be the reason for their not having any lice. However, even mammals that spend most of their time in the sea may have lice, if they come to land at some time during the year. (Lice only mate when exposed to the air.) This is the case with the Pinnipedia (sea lions, walruses, and seals). As for the Pholidota (pangolins), their lack of lice may stem from their lack of hair and the toughness of their skin, or it may be due to their close association with ants. Anteaters are also lice-free, and since it is known that the formic acid some ants produce is lethal to lice, it might be that feeding on ants and living in their nests is both good and good for you. Tough skin and lack of hair are also seen in the Proboscidea (the elephants), but in that case there is a single mallophagan louse that has managed to evolve along with its host and keep pace with the thickening skin.

When several different species of Mallophaga find themselves on a single host bird, interspecific competition forces a segregation that in the end minimizes competition. In this way, these species embark on different evolutionary paths. For instance, the Mallophaga that inhabit the head of a bird normally have larger heads and stronger mandibles than, and are ponderous when compared with, the "lowland" species, which are normally more flattened and fleet-footed in order to escape the bird's beak when it preens.

Some Mallophaga have adopted extremely specialized habits. We have already met the pelican's louse, with its inside-the-pouch existence, and then there is the strange case of *Actornithophilus patellatus*. This louse spends its whole life inside the quills of the primary and secondary feathers of the curlew. Here it eats the pith, copulates, lays eggs, and dies. The only reason it may leave a feather is if conditions become too crowded; then it goes in search of greener pastures a few centimeters away.

The few lice that parasitize pinnipeds have to contend with cold, even frigid water most of the time. For example, the Weddell seal inhabits antarctic waters and often comes out onto the ice to rest. The Weddell seal's louse (*Antarctophthirus ogmorhini*) has been shown to survive a temperature of -20°C for 36 hours, more than enough to protect it during the seal's cold respites at the surface. This louse is prepared to mate and reproduce whenever the temperature climbs above 5°C. Another louse, *Lepidophthirus macrorhini*, lives on the southern elephant seal. It comes to shore, on its host, twice a year. During the egg-laying period, the females go into a frenzy, laying up to nine eggs a day for a period of four or five weeks. Like many gilless aquatic insects, both these lice are covered by small flat scales that act as a plastron retaining air over the spiracles—their own self-contained diving apparatus.

Like in most of the rest of classical biology, a few biological rules have been devised to summarize knowledge of parasites and their hosts. Lice illustrate some of these rules quite well. For example, Fahrenholtz' rule says: "In groups of permanent parasites the classification of the parasites usually corresponds directly with the natural relationships of the hosts." As is the case with most biological rules, examples can be found that verify this rule, but exceptions that test it can also be found. For instance, the kiwi's louse is related to those that infest rails, even though the kiwi was at one time thought to be related to ratites (ostriches and their kin). However, there is now some anatomical evidence that the kiwi is related to rails. An example that verifies the rule is that of the genus of sucking lice that parasitizes humans; it also afflicts chimpanzees, and a related genus afflicts Old World (cercopithecoid) monkeys. In contrast, flamingos, though anatomically closer to storks than to ducks, share several of their ectoparasite genera with ducks.

Since lice are so dependent on their hosts, it is reasonable to assume that when the host dies, the entire population of lice living on that host perishes. Sometimes this is not the case. A dead host becomes quite disagreeable to lice, and they will try to leave it as soon as possible. In some instances, when the host is part of a herd or flock, the lice have a chance of reaching a new host, but only if that new host brushes against the dead individual. In a few cases, however, biting lice have commandeered the bodies of other more mobile parasites in order to leave the proverbial sinking ship. A well-documented case is that of two parasites of starlings—the mallophagan *Sturnidoecus sturni*, and the fly *Ornithomya fringillina*. In one recorded sampling of starlings and their parasites, almost half the flies had lice attached to their abdomens, and one fly was carrying 22 passengers. These flies have apparently become a taxi service for the lice, and even though the flies are not confined to just starlings, eventually a fly will land on another starling and the lice can get off. *Ornithomya fringillina* belongs to the dipteran group pupipara, whose members do not lay eggs, but instead produce a full-grown or almost full-grown larva ready to pupate. Other rarer cases of phoresy (as this curious habit is called) have been reported, including the transport of human lice by house flies.

Earlier in this chapter we saw that some biting lice derive much needed water from wounds on their host's skin, but this is an episodic and random source of moisture; another such source is rainwater. Some biting lice have tapped into a much more reliable source of moisture: the water vapor available in the lice's immediate atmosphere; that is to say, in the thin layer of still air trapped among the feathers or fur of their host.

The Mallophaga are the only other insects known to have honed water-vapor absorption to the level attained by the thysanurans and the death-watch beetles. The water-vapor-absorption apparatus of mallophagans is completely different from the others previously described, but it is similar to the water-vapor-extracting apparatus of the book lice and lends support to the idea that biting lice evolved from a psocid-like ancestor. The apparatus consists of a pair of eversible bladders of the ventral hypopharynx that the insect places into position by opening its mouth. A pair of glands associated with these bladders secrete a hygroscopic liquid over the bladders as the first step in absorption. The surface film on the bladders is connected through a small capillary tube with a powerful piston-like pump located on the floor of the insect's mouth. Operation of this pump draws the condensed water on the bladders through the capillary tube and eventually delivers it to the gut. The rapidity of the pumping action determines how much water the insect can collect, but mallophagans are able to pump at a much faster rate than book lice. A dehydrated louse will pump faster than a hydrated one, but if the relative humidity becomes too low, the insect will shut the system down, since at that point it risks losing moisture. If the relative humidity falls below 43%, these insects stop pumping, retract the bladders, and seal the preoral cavity.

It is clear from the vastly differing designs of the various water-vapor-extracting apparatuses described in this chapter that the trait has evolved independently several times, and there might be other designs out there just waiting to be discovered by some entomologist with an inquisitive mind and the patience of Job.

Loose Ends: Canthariasis

Several other groups of arthropods enter vertebrates' bodies either purposefully or accidentally. In a few instances it is easy to find a series of intermediate stages from totally free-living species to those that are parasitic. Perhaps the behavior of the scarab beetles is the easiest to understand. As a family, these beetles make it their business to use fresh dung as the source of nourishment for their larvae. There is stiff competition among dung beetles for this resource, and since old dung is hard to mold, the fresher the dung, the better. In some regions competition has gotten so stiff, and dung has become such a prized commodity, that the adult beetles have gone directly to the source. They climb on the ungulate's hind quarters and wait for the next fresh patty to ride to the ground. Some beetles are more impatient than others; their impatience makes them enter their victim's anus to take their quarry. In India and Sri Lanka, some species of *Onthophagus* take this last step; unfortunately, they do not discriminate in their choice of victims and often target children.

Other beetles do not enter the human body purposely but are sometimes ingested with contaminated food. Some of these are resilient enough to survive a passage through the human intestine. These include the yellow mealworm (*Tenebrio molitor*), so often used to feed pet lizards, the black carpet beetle (*Attagenus megatoma*), which used to be much more common in the days before synthetic carpet materials, and the spider beetle

(*Ptinus tectus*), so named because of its long legs and compact, shiny black body. Other such unwelcomed invaders of human cavities include several species of centipedes that have been found infesting human nasal, intestinal, and urinary tracts. Some species of millipedes (mainly in the genera *Julus* and *Polydesmus*) have also been reported infesting the digestive and urinary tracts of humans. Besides causing discomfort, some of these animals can also be vectors of helminth diseases. For instance, mealworms (*Tenebrio* spp.), centipedes, and millipedes can be intermediary hosts for the dwarf tapeworm (*Hymenolepis nana*). Even though this parasitic infection is rare in the United States, its seriousness makes ingestion of these arthropods a serious matter.

True Worms (Helminths)

We will concentrate on the life history and physiology of the dwarf tapeworm, not only because it is one of the most common intestinal parasites of humans, but also because it is remarkable, even among tapeworms, in that it may complete its life cycle entirely within its primary host. This tapeworm is only from 7 to 100 mm long as an adult and is among the smallest of the mammalian tapeworms. The larger the number of worms present in one person, the smaller their average size. This correlation is most likely a response to heightened competition among the worms. Infection happens through ingestion of the embryonated eggs present in human excrement; for this reason this tapeworm is most commonly found in children who have not yet developed toilet habits. The ingested eggs hatch in the upper small intestine and the larvae that are released penetrate the villi—the small finger-like projections of the intestinal wall. Here they stay until they mature into the next larval stage, at which time they emerge back into the lumen of the small intestine. They then attach to the mucosa and, in a few weeks, transform into adult worms.

The previous sequence of events is the usual life cycle for these worms, other variations to the life cycle are possible, however. As we have already mentioned, sometimes the immature stages are found in beetles or millipedes, and cases of infection from these sources have been documented, but they are rare. This worm can also undergo a life cycle variation known as autoinfection. In this case, a very high population of worms can develop because the eggs manage to hatch in the intestine without the need to travel to the outside world. It has been shown that such infections are possible in immune-compromised laboratory animals, and the heavy worm burden these infections cause can lead to health complications.

Dwarf tapeworms, as well as other helminths, have completed the transition to parasites, and cannot exist outside their hosts, except fleetingly. The conditions that make the intestine of large organisms a limiting environment are the lack of oxygen and high carbon dioxide levels. These conditions are similar to those in the anaerobic sediments we will encounter in wetlands, except that here there is no chance to make it to the surface for a gulp of air. Helminths deal with these conditions by utilizing a set of pathways that allows them to use the plentiful carbon dioxide to dispose of electrons. Here this technique is taken to an extreme not encountered in anaerobic mud. The worms we are discussing now do not expect to make any trips to more oxygenated zones; the intestine is their home, and here they will stay for the rest of their days. The set of pathways that allows them to derive energy and balance reducing equivalents depend on the ability to coferment glucose and amino acids. This ability is not unique to worms, and mollusks that spend many hours tightly sealed during low tide also possess similar pathways. However, the fine-tuning that helminths have achieved has allowed some to spend most of their life cycle without oxygen.

4 From Sand to Serpentine

Dunes

Few of the geological processes that impact life on this planet proceed at rates comparable to the rates of growth of most living things. The processes that create sand dunes are some of these few, and sand dunes react to the environment with the speed of a living organism. Because the motion of the dunes is on a time scale similar to that of plant growth and development, the survival of a plant growing on such a dune becomes a race, in which a living being is pitted against a nonliving, but very animated object. This race is often lost by the plant; for instance, only a few plants can colonize the shifting sands of a beach dune system.

Other factors, besides their ever-changing surface, make dunes particularly harsh for most plants. Two of these factors are a dune's special water-holding character, and the low nutrient levels often found in well-sorted sands. These factors conspire to make dunes extreme environments for plants and animals, unless these organisms possess special adaptations. It is not that dunes are deserts; on the contrary, in true deserts dune fields exemplify places where plants and animals have ready access to water—but only if they have those special adaptations. Water is held deep within the dune system, away from the drying forces of wind and sun. If the water table is found three to four meters deep in a sand dune, it may seem that the only prudent course for a plant would be to produce a deep taproot, and for an animal, to dig a deep burrow. However, some dune-colonizing grasses do not extend their roots deeper than a meter, and many animals do not dig deep burrows; they all rely on rainfall and on the "internal dew" of the dune. In many instances, the animals of sand dunes rely on the plants to extract moisture from the store inside the dune, and they simply harvest this bounty in the form of sap, leaves, roots, or seeds. We have already seen that a good pair of kidneys go a long way toward making such a source of moisture all that a desert rodent requires. For this reason, in this section we will concentrate on the special adaptations of plants.

Sand dunes have special properties that make them distinct from their surroundings no matter where they are located. We recognize a sand dune as a distinct entity whether in the Sahara in Africa or along the Kobuk River in Alaska. The species of plants that inhabit dunes in these two areas are completely different, even to the extent that only two families have members on both sand-dune systems. Yet, there is a special intertwining between the environmental factors that affect sand dunes, wherever they may be. This intertwining causes disparate species belonging to different families to evolve along parallel tracts, as if striving toward an ideal Dune Plant. There are a few complications in this simple picture, however. One of these complications is the diversity of dunes themselves. There are several different kinds of dunes, each created by a different set of conditions, and we must first discuss some of the factors that create dunes and contribute to their harshness before we can deal with the physical dune types. Only then can we reach an understanding of the driving forces molding animals and plants on dune fields.

Dunes are children of the wind. They are born from several factors: climatic factors that determine a predominant wind direction, a propitious positioning of a source of fine particulate, and topography. If precipitation is sufficient, then only very strong winds or high-energy surf will maintain an active dune system against the encroachment of vegetation. In drier areas the wind or the surf can be less energetic, and significant dune systems may still form, and in turn these dunes will encroach on the surrounding vegetation.

Ralph Bagnold, a British officer serving in Egypt during the early part of this century, was the first European to delve into the mysteries of dune formation. His tour of duty in Egypt exposed him to some of the most active dune fields in the world. His curiosity piqued, Bagnold returned to England to pursue experimental studies in the processes of dune formation. He had observed that some dunes are crescent-shaped, while others form parallel ridges. The crescent-shaped dunes or barchans are formed in steady-wind conditions where there is a moderate supply of sand. Under such circumstances, almost any obstruction will cause a dune to form, and the sides of the dune parallel to the wind direction will be drawn out as crescent-shaped arms in the direction of the wind. Where the wind shifts direction constantly, and the sand supply is limited, star-shaped dunes will form, and in some places, such as the Dune Seas of the Namib, such star dunes may be over 300 meters high.

Sand dunes can be like the mythical phoenix, but on a geological time scale. They may have lived, then died and been buried in some shallow sea where carbonate cement and pressure may have fused the sand grains into stone. Later, this sandstone may have been exposed again and weathered, only to revert to sand. In the appropriate setting, this sand may again be picked up by the wind, sorted, and deposited in some convenient place to again form a dune. This phenomenon can be seen in places such as the Coral Pink Sand Dunes State Park of southern Utah, where Navajo sandstone has reverted to sand. But sand, by itself, is not enough; there must also be some geographical features that make the wind pick up the sand grains, push or cajole them along a common path, and then release them in unison. At the

Coral Pink Sand Dunes, there is the fortuitous convergence of the Markagunt and Paunsaugunt Plateaus. The steady winds are forced down the natural venturi formed by the Sevier fault, and as this happens, they drop their load of rusty sand.

Beach Dunes

Beach dunes owe their formation and continued existence to the onshore wind that is a shaping factor of almost every seashore. This wind is a local phenomenon generated by the difference in heating rates between the land and a large body of water. It takes a lot of heat to raise the temperature of water one degree Celsius; we call this value *the heat capacity*. Among naturally occurring materials, water has one of the highest heat capacities. Consequently, the same amount of sunshine falling on and being absorbed by, a body of water and the adjacent land, will raise the temperature of the land more than that of the water. Since the land warms up faster, the air over the land also warms up faster than the air over the water. This difference generates an upward flow of warm air over the land, and a downward flow of cooler air over the water. The wind thus generated during the day is that wonderful sea breeze that lures families to the beach on warm summer days. It is also the steady wind that helps create the dune systems of sandy beaches. Whether the dunes are tall (12 meters along some stretches of the Gulf of Mexico), or barely present at all, depends on certain factors. Among these factors are the longshore supply of sand and larger-scale winds; for example, those associated with the cold fronts that, mainly in fall and winter, penetrate down to the coasts of the Gulf of Mexico, bringing northwesterly winds. On those coasts, the great destroyers of primary dunes (besides man's lust for beachfront property) are the tropical storms of summer and fall.

A little beach topography is in order at this point. The zone between the lowest low tide and the spring tide is known as *the foreshore*; it is an area devoid of vascular plants but not devoid of life, and many minuscule creatures—including both plants and animals—make the interstices between the sand grains their home. The turmoil of this zone makes it very difficult for vascular plant propagules to colonize this environment. The zone between the spring tide and the base of the primary dune is *the backshore*. Next come the dunes closest to the beach proper, the primary dunes. They are made up of almost pure sand with no water-retentive organic material. Salt spray, sand blasting, unrelenting sunshine, and the lack of organic matter make the colonization of the sand surface on the windward side of the primary dunes a daunting task for even the most tenacious plant. However, a few hardy pioneers can grow fast enough not to be completely buried or left totally exposed by the shifting sand.

Among the most perplexing mysteries of the dune system is its extreme dryness. Coarse sand is very porous and, if it has been newly deposited by the wind, it contains little organic matter. Water in such sand percolates rapidly and is soon out of the reach of shallow-rooted plants. The roots of dune pioneers such as sea oats and beachgrass penetrate only a meter or so into the sand. Since many dunes are taller than the depth of root penetration, these plants are not tapping the water table under the dunes directly. Could capillary action be raising the water level high enough for the plants to avail themselves of this water? This proposal has been tested, and it was found that even in very fine sand, where capillary action should be maximal, the rise in the water level was no more than 40 cm. Most of the time, pioneer dune plants obtain their water from rainfall. Although most of the rain water penetrates deeply, some of it lingers around the plant roots, as pendular water, long enough for the plants to imbibe it and survive. Many of these plants survive periods of drought, so there must be some other source of water available to them. One possible means by which the water table could contribute to a plant's moisture supply during periods of drought is through an exchange with the water vapor in the pore space of the dune. As night falls, a front of dropping temperatures progressively penetrates into the dune from the surface. As morning approaches, the surface is again beginning to warm, but the cooling front of the previous night is still progressing downward, although its effect is much decreased. Because of the time it takes the front to reach a depth of about 35 cm, the cycles of warming and cooling at this depth are reversed compared with those experienced at the surface. The interstitial air at depths below 90 cm is saturated with water vapor. During the following day, as the cooling front moves from 35 cm progressively down toward 90 cm, water vapor moves upward toward the surface. At around 35 cm the lower temperatures rob the water molecules of their energy and cause them to settle within the root zone of the plants.

We will start our survey of dune systems from the north polar regions and work our way south. Of the many dune systems all around the world, only a few have been studied in any detail; with some exceptions, they are mainly dune systems in easily accessible areas in subtropical climates.

Survey of Dune Systems

In Alaska there are extensive sand dunes on the arctic slope south of Barrow. These dunes are among the largest sand-dune systems in North

America, stretching from the Meade to the Colville rivers, but they have been studied little. The coastal dunes along the Seward Peninsula north of Nome have been studied more. The dune formations in this area are controlled by the onshore wind and do not reflect regional wind patterns. Small barchans less than a meter high are present on the widest beaches.

On the Bering Sea coast of the Seward Peninsula there are oval blowouts (depressions made by the wind in the sand). These blowouts are dotted by miniature one-meter-high dunes and smaller ridges. The main colonizer of these minidunes is lymegrass (*Elymus arenarius*); the ridges are covered on their leeward side by a dryad-sandwort turf. The windward side of the ridges is almost vertical and actively ablating, thus exposing the roots of the plants growing on the other side.

Farther east, along the Kobuk River, aeolian sand deposits form an extensive dune field that is now part of the Great Kobuk Sand Dunes National Park. Because of its park status, this dune field is the most intensively studied dune system in Alaska. There are several factors that affect the evolution of these dunes from unconsolidated mounds of sand to well-vegetated, stable dunes. These factors include the instability of the substrate, the lack of nutrients in the sand, and the changing water availability inside the dune. The initial instability of the substrate prevents colonization by plants whose strategy for water-and-nutrient retrieval consists of producing long taproots. Such plants cannot become established on dunes that are actively moving, since in those circumstances their slow above-ground growth would be quickly buried. The plants that colonize such dunes the world over are plants with stoloniferous growth, with the ability to quickly root at the nodes and the ability to elongate the internodes to keep pace with the accumulating sand. In the Kobuk area these plants are a sedge, *Kobresia myosuroides*, and the red fescue (*Festuca rubra*). Along with these, a few taprooted plants manage to grow; these include the arctic sandwort (*Minuartia arctica*) and a milk-vetch, *Oxytropis kobukensis*. The oxytrope, a legume, has an initial advantage in the nutrient-poor dune system (as long as the soil is not completely devoid of molybdenum), since it has symbiotic bacteria that change atmospheric nitrogen into a form usable by the plant. All these pioneer dune plants stabilize the dunes and provide a stable surface that allows lichens to colonize. Some lichens fix nitrogen also, and some form a crust on the surface of the sand, thus limiting the exchange of water vapor between the dunes and the atmosphere. These changed conditions allow slow-growing plants to colonize, at the expense of the grasses and sedges. Those other plants include the moss campion (*Silene acaulis*), other sandworts, and rock jasmine (*Androsace chamaejasme*). At this point the sand dunes can be over 60% vegetated and have extensive lichen crusts.

In the polar and temperate zones the dune pioneers are grasses. (In the tropics, diversity increases, and though grasses are still present, they are no longer dominant.) These grasses have a few important traits in common: (1) They are rhizomatous plants able to spread through underground stems. (2) Their root systems do not extend more than a meter deep. (3) They are extremely salt-spray tolerant. In arctic North America, in those protected places that are not scoured clean each winter by sea ice, lymegrass is the most common dune pioneer. A related species, dune wildrye (*Elymus mollis*) used to be the dominant species along the Pacific coasts of temperate North America before the introduction of the European beachgrass (*Ammophila arenaria*). In Newfoundland, the dominant species is American beachgrass (*Ammophila breviligulata*), which ranges south to Virginia. Still farther south along the Atlantic beaches of North America, the American beachgrass is replaced by sea oats (*Uniola paniculata*), which is dominant on many sandy beaches throughout the Gulf of Mexico and is also found (though it is not dominant) well into Central America.

Gulf Coast Beaches

The first sign of vascular-plant life is normally found where the spring tides leave their load of debris, which includes seeds of ubiquitous beach pioneers such as the sea rocket (*Cakile edentula*), the sea purslane (*Sesuvium portulacastrum*), and the seashore-elder (*Iva imbricata*). Even though these three succulents can sometimes be found growing farther toward the surf than any other plants, they seldom cover large areas. The seashore-elder does contribute to dune building. Its semierect stems root as the sand accumulates around them, and small mounds of the plant, up to a meter high, can sometimes be found where the rate of sand accretion matches the plant's rate of growth.

Other pioneers that make the area between the spring-tide mark and the base of the primary dunes their home are several vines; they include the cosmopolitan morning glories (*Ipomoea* spp.), such as the railroad vine (*Ipomoea pes-caprae*) and the white morning glory (*Ipomoea stolonifera*). Traveling south along the coast of the Gulf of Mexico, the beach ecosystems grade from temperate assemblages of species along the northern Gulf coast, to tropical ones as one travels through southern Texas and thence to northern Mexico. As this transition occurs, grasses become less prominent, and shrubs belonging to several families become the dominant vegetation along the Yucatán coast and on the Caribbean Islands.

Along the northern Gulf coast, the dunes are colonized by several rhizomatous perennial grasses; the most attractive of them are the sea oats, with their large flat panicles that ripen in fall and last through winter. Many times sea oats are dominant along the crest and windward parts of the primary dune system, although sometimes other grasses also make a strong showing. These include the robust bitter panicum (*Panicum amarum*) and the more dainty seacoast bluestem (*Schizachyrium maritimum*), which is more at home on the lee side of the dunes. The sand dunes of the northern Gulf coast are at their best development along a stretch from the Apalachicola River to the Alabama River. Here we find several indigenous species such as the Florida rosemary (*Ceratiola ericoides*), which is common on the leeward side of older secondary and tertiary dunes, and a sand square, *Paronychia erecta*, which is common on the leeward side of the primary dunes and also along the wet interdune swales where the water table is near the surface. Along the windward side and along the crest of the primary dunes we also find unexpected residents like the devil-joint (*Opuntia pusilla*) and the more common prickly pear (*Opuntia humifusa*).

Along the whole length of the Gulf coast, the most common animal on the dunes is the ghost crab (*Ocypode quadrata*). This little buff-colored crab is a generalist and a scavenger, perfectly suited to the variability of its environment. If coquina clams (*Donax* spp.) and mole crabs (*Emerita portoricensis*) are abundant, the ghost crab will hunt for them at night. On other beaches, this crab will feed on plant or animal matter that washes ashore. Human refuse, including areas of spilt liquids, is very attractive to these crabs, and large numbers can be found at night sifting the sand where someone spilt a drink the previous day.

As we travel south along the Gulf coast of Florida or Texas, we encounter a much more imposing crab, the red land crab (*Gecarcinus lateralis*), which can attain a carapace width of 10 cm. (This crab is also found in the Caribbean region, where it is one of two commonly encountered land crabs.) In southern Texas and southern Florida, the red land crab is at the northern limit of its distribution, for winter cold is incompatible with its chosen mode of respiration. This crab is significantly more independent of the sea than the ghost crab. Its burrows do not need to reach free water, because it has along its legs special setae (hairs) that siphon water from merely moist sand or vegetation into its gill chamber, thus keeping its gills moist and allowing it to breathe. For this nocturnal crab, this anatomical design works well in warm climates, where it may be active every night of the year and can replenish its gill water from the dew that settles on surfaces. In climates in which there might be freezing spells lasting over a few nights in winter, surfaces collect frost and not dew at night (when the temperature is too cold for this crab to venture out of its burrow anyway), and during the day the humidity drops substantially. Under such conditions this animal soon suffocates.

Soils of Volcanic Origin

Sand is not the only substrate that poses difficulties for plant growth; newly erupted volcanic material can be even more inhospitable. Volcanoes come in all shapes and sizes, from the giant volcanoes whose tops we know as the Hawaiian Islands to the small cones that dot the Utah-Arizona border in the southwestern United States. Volcanoes also come in all kinds of flavors; there are wild andesitic volcanoes like Mount St. Helens, and there are more serene basaltic volcanoes like those that formed the Hawaiian Islands. Since a volcano is a manifestation of processes that go on deep underground, some volcanoes can and do change their temperament over time. Such a transformation depends on the makeup of the material that is bubbling toward the surface at a particular geologic moment. Andesitic volcanoes, because of their unconsolidated surfaces composed mainly of ash and pumice, are easily worn by the action of wind, water, and frost; and they eventually become an environment that is hospitable to life. A perfect example of this is Mount St. Helens; ten years after the 1980 eruption the surface was literally covered by fireweed and tree seedlings that had started germinating. In stark contrast, several basaltic volcanoes that we will survey from Iceland to the Hawaiian Islands will show how difficult the hard angular basalt is to conquer under some circumstances. (However, the soils that are eventually produced from basalt can have higher concentrations of plant nutrients than rhyolite-derived soils.) Part of the difficulty is the lack of any water-retentive materials in newly erupted lava. Sand can hold some water in its pore space, but the cracks in lava let the water percolate very deeply before any plant has a chance to capture it. A second problem is the dark color of fresh (less than a few thousand years old) basaltic lava. The dark surface causes the lava to heat to levels that are lethal for seedlings.

The Hawaiian Islands are the surface manifestation of a “hot spot” under the Earth's mantle. The two tallest volcanoes of these islands are Mauna Kea and Mauna Loa; because of their midoceanic location, the lava these shield volcanoes produce is basaltic and has a low silica content, which makes the eruptions of these volcanoes much less violent than those of andesitic volcanoes. Mauna Kea is older than Mauna Loa, and its age is reflected in its vegetation. Mauna Kea has vascular plants growing near the summit, whereas Mauna Loa scarcely has any plants at all above 3400 meters. It is not often realized that a dark object is not only a good absorber of visible (and usually also infrared) radiation, but also an equally good emitter of the same radiation. A black lava surface is a good absorber of long-wavelength radiation, and by this rule, it is also a good emitter. During the day, this surface absorbs sunlight and heat, and the temperature builds. As the surface warms up, it emits progressively more energy as long-wavelength radiation. The air above the lava, on the other hand, is a poor absorber and emitter of both short- and long-wavelength radiation; it simply lets radiation through. If this dark lava surface happens to be at the top of a tall volcano for instance, the lava will cool down below the air temperature through skyward radiation during the night. At high altitude a seedling on lava might bake during the day and freeze at night.

The difference in the ability of the two types of lava produced by such basaltic volcanoes to support life is also striking. On Mauna Loa, aa (jagged) lava is completely bare above 3050 meters, while pahoehoe (ropy) lava is colonized, at least by the hoary rock moss (*Racomitrium lanuginosum*) up to the summit. (This cosmopolitan moss is even found in the the Arctic, where it forms large mounds.) In the pahoehoe lava this moss survives as little wisps stuck in cracks where moisture stays a little longer in the morning. Another telltale sign of the difficulty of colonizing fresh lava is the dearth of foreign weeds on the slopes of Mauna Loa. Whereas Mauna Kea has several cosmopolitan weeds ascending its flanks up to about 3700 meters, Mauna Loa's weeds do not ascend past 2500 meters. Eventually, Mauna Loa's lava fields will transform into rich nurturing soils, as happens to low altitude lava fields on these islands. But on Mauna Loa's high slopes, with meager precipitation and organic deposition, that prospect is in the distant future.

Expansive Clay Soils

There are other volcanic-derived soils that pose particular hazards to plant life. Soils derived from volcanic ash can, under some circumstances, have high amounts of expansive clays; such soils can be found in many arid and semi-arid locations, and many times they are almost devoid of plant life whereas nearby soils might have a reasonable plant cover. Bentonite is the name given to a complex expansive clay from western North America. This clay consists mainly of montmorillonite, which for plants is a nightmare. I have seen sagebrushes be undermined by every rain shower; they had had the misfortune to germinate on bentonite hillocks in Dinosaur Provincial Park in southern Alberta. By the time I happened by, the plants were barely holding on by the tips of their taproots (which had been deep within the soil not long before), most of them toppling. One or two sagebrushes had managed to outlast their toppled comrades and were still standing erect holding on at the top of most of the hills. What makes this clay such a difficult substrate is its quaint reaction to water. The sides of a bentonite hill can be covered (depending on the amount of precipitation) by what looks like gray popcorn. This material is what is left of the top of the hill. As rain falls, you hear a cacophony of popping sounds. These sounds are not caused by the patter of rain; they are made by the clay itself! You expect such sounds from a breakfast cereal, not emanating from the earth. Bentonite expands as raindrops wet the top of the hill, and like popping kernels of corn, the little pieces of surface material take odd rounded shapes and roll downhill, gradually exposing the plant's root system. Only plants that can modify the morphology of their root systems depending on the situation (the sagebrushes, for example) can have even the slightest chance of surviving such a fickle surface. In even drier regions, this clay makes impervious mounds that are almost totally devoid of higher plant life. It is, however, a relatively benign substrate for microbial growth since it loses water slowly to the atmosphere.

Surtsey

On November 14, 1963, in the cold waters southwest of Iceland, an underwater volcano erupted violently. Its first eruption spewed ash, cinders, and pillow lava. This eruption, and the ones that followed, created a submerged mesa that served as the base of what eventually would be known as the island of Surtsey. The eruptions continued over the next four years, piling fluid basalt over the cinders until the island rose 173 meters above the waves, and on the fourth of June, 1967, the eruptions stopped. Surtsey (named after Surtr, the Norse god of fire) is the only survivor of three islands that were thus created; it is 2.7 square kilometers and lies about thirty kilometers southwest of Iceland.

Surtsey's first inhabitant of any kind was a chironomid midge, *Diamesa zernyi*, which must have arrived a few months after the island's birth. By the fall of 1964 there were also two moth genera, including the ypsilon dart (*Agrotis ypsilon*) so common in North America, one mite, and six species of flies. Just a few years later, the number of animal species detected on the island had risen to seventy. As is the case for the whole Arctic, most of these arthropods were flies; in all, there were forty-three species. These numbers are deceiving, however, since only a small fraction of the arthropod species managed to found thriving colonies on the island.

While the mostly aerial invasion by invertebrates was going on, the flowering plants were barely beginning their colonizing sea voyages. Some plants did arrive quickly by air, but these were not mineral-soil pioneers and did not germinate. The tufted seeds of the common groundsel (*Senecio vulgaris*) can travel many kilometers carried by the prevailing winds, and did arrive on Surtsey early on, but the harshness of the newly emerged basaltic surface made the voyage futile. The black volcanic ash of this infant island is high in mineral salts and is completely lacking in organic matter. The black basaltic surface of the interior of the island gets very hot under the unrelenting sunlight and holds no moisture, so the best location for plant survival is on the beach. Since the island's birth in 1963, a steady progression of plants has invaded its shores. The plants that are best suited to colonize newly emerged shores are those backshore pioneering species that have adapted to use sea currents for their long-range dispersal. Their seeds are not immediately killed after immersion in salt water and thus might make it to a distant shore. These plants are also adapted to extract nutrients from humus-free mineral substrates. Of these pioneering plant species, the first that arrived at Surtsey's shores, in the summer of 1965, were the sea rockets (*Cakile* spp.)—annual plants that pioneer the drift line on beaches throughout the northern hemisphere. By 1967 they had been joined by lymegrass, along with sea-beach sandwort (*Honkenya peploides*) and sea lungwort (*Mertensia maritima*). Land plants were much slower to get established; by 1967 only mosses had found footholds close to the older vents, which were still geothermally active. No lichens, normally thought of as the first pioneers, had yet come to Surtsey.

The populations of sea rocket and lymegrass crashed in the winter of 1973-74, and both had become extinct by 1981. New lymegrass seedlings appeared in 1983, and the population started a rapid expansion. The sea rocket, however, had yet to reappear by 1986. In the winter of 1974-75, the sea-beach sandwort and the sea lungwort were the ones that suffered dramatic losses; whereas the sandwort quickly rebounded, it took the lungwort six years to start increasing in numbers again. By 1986 the most numerous colonizer by far was the sea-beach sandwort, with hundreds of thousands of individuals, while all the other plants amounted to a few hundred individuals. The sea-beach sandwort can grow in the ash and pumice on the beach, and each plant forms a small mound of branched rooting stems on the unconsolidated substrate with a top dressing of windblown sand. After three to four years, these plants mature and start releasing thousands of fertile seeds, and a population explosion can ensue. On Surtsey, the lymegrass has taken advantage of the benign conditions inside mats of sea-beach sandwort, and many lymegrass seedlings can now be found growing in such mats.

Since the plant populations account for little biomass, herbivorous animals have found it difficult to colonize Surtsey. Most of the species that have taken up permanent residence on the island are detrivores or scavengers, and a heleomyzid fly, *Heleomyza borealis*, is the most abundant colonizer. This fly depends on carcasses of birds or fish that wash up on the beach, and it must compete for these tidbits with the gulls that frequent the island. Another colonizer, a chironomid midge, *Cricotopus variabilis*, makes its home in tidal pools and feeds on algae that grow in these pools. Another successful insect on Surtsey is a springtail, *Archistoma besselsi*, which, like many other collembolans, can ride the ocean currents on the surface as if they were automated walkways. (This collembolan is a common member of the fauna found among the grains of sand on beaches throughout the north temperate zone.) On Surtsey, away from the beach and in the moist moss patches around active fumaroles, only a few microscopic amoebae, ciliates, and rotifers may be found.

Serpentine and Other Heavy-Metal Soils

Most plants need soil for support, as a source of nutrients, and as a reservoir for water. Soil, however, can be laced with toxic quantities of metals or other elements, whether naturally or because of anthropogenic (man-generated) contamination. Few plants can survive the rigors of these mineral soils. Of the ones that do survive, some exclude the toxic elements at the level of the roots. Others, in what seems a suicidal revelry, accumulate the toxic substances many fold over the level found in the soil. Yet they do not succumb. Although resistance to the toxic elements can evolve quickly (over one or two generations), the strange accumulator habit, surely a perverse addiction, probably takes a very long time to evolve. These interesting plants deserve more study than they have received, not only as beautiful examples of evolution in progress, but also as examples that might shed light on how we may deal with the toxic areas created by our mining and smelting activities.

In this section we will discuss toxic soils of several kinds. We will start with a particularly toxic type of soil called *serpentine*. Serpentine soils are the weathered remains of intrusions that originated deep within the earth. Their composition is full of dense elements worthy of the forges of Hephaestus. Because of their alien nature, serpentine soils place extreme selection pressures on plants, and areas containing such soils can seem desolate in comparison to nearby areas with nontoxic soils.

Serpentine soils' heavy-metal content is not the only reason they are toxic. Other reasons include their very high Mg/Ca ratio—which inhibits a plant's ability to take up Ca—and their low nitrogen, phosphorus, molybdenum, and potassium content. Plants that grow on serpentine adapt to a high Mg/Ca ratio by decreasing the efficiency of their magnesium-uptake mechanism. This adaptation causes them to suffer from magnesium deficiency when they grow in normal soils. From this description it may seem that such soils would be more hospitable were they to be fertilized. In general, this is not the case.

Fertilization experiments carried out with serpentine-adapted grasses showed that raising nutrient levels did not significantly enhance growth; on the contrary, it destroyed the plants' tolerance to the toxic materials. Nitrogen fertilization was ineffective on magnesium soils because the nitrogen source did not affect the Mg/Ca ratio. Nitrogen fertilization was also ineffective on nickel soils because the ammonium used as a nitrogen source increased the acidity of the soil, and as a result, there was an increase in heavy-metal uptake by the plants and therefore in the toxicity to them. When nitrate was used as the nitrogen source instead of ammonium, it stimulated the plant's uptake of iron and aluminum, even up to toxic levels. High fertilizer phosphorus levels also produced problems, and the toxicity of nickel was greater in the presence of high phosphorous and low calcium and potassium levels.The best treatment for increasing plant growth proved to be calcium fertilization. All these experiments were carried out on grasses, and thus may not apply to other types of plants.

There are several extensive serpentine areas around the world, and each of them is more desolate than the surrounding areas of normal soil. Their barrenness is often exacerbated by mining activities. Most of these areas have been exploited for their mineral worth, but only a few have been scientifically studied to any extent. They are the sites of extensive surface mines where nickel, cobalt, and copper are the main products. Because of that mining, some species of accumulator plants (which are endemic to these areas) are vanishing, since they depend on metalliferous soils and usually grow in only one locale. This is the case at Dikuluwe and Mupine, west of Kolwezi, Zaïre, where an endemic three-seeded mercury, *Acalypha dikuluwensis*, an endemic love grass, *Eragrostis dikuluwensis*, and other accumulator plants were once easily found. If these vanishing species are not rescued soon, their potential for mine reclamation may never be known. The same mineralogical conditions that led to the evolution of these amazing plants are now leading to their downfall at the hands of man.

The accumulation of toxic elements by plants is a well-established phenomenon. Most of the soil-contaminating metals that modern industrialized societies produce (some of these are shown in table 1) have stimulated the evolution of accumulator plants. Often, such plants can be found on or around active, as well as abandoned, mines and smelting operations. The amount of time required to produce a species or variety that is tolerant to high levels of a particular metal is so short (one or two generations) that it may be observed experimentally. In contrast, a much longer time is required to produce a plant that is dependent on this metal for its survival. When such a plant evolves, it becomes a good *geobotanical indicator* of the presence of the particular element, since it grows on sites contaminated by the element and nowhere else. When a plant is an indicator

Table 1 Typical industrial sources of heavy-metal pollution

	Metals in Waste Stream												
Industry	Ag	As	Cd	Co	Cr	Cu	Fe	Hg	Mn	Ni	Pb	V	Zn
Smelter Coke		●				●							
Coke Ovens				●							●		
Cu Smelters		●				●	●						●
Zn Pipeworks													●
Pb Smelters	●	●	●								●		●
Zn (Pb) Smelters			●								●		●
Coal Burning Powerplants			●				●			●			●
Cooling Towers					●								
Metallurgical works											●		●
Oil-fired powerplants			●		●	●		●	●	●	●	●	●
Zn Smelters			●			●			●	●	●		●
Sewage Sludge Incinerators											●		
Saw Mills		●											
Chlor-alkali works								●					
Ni Cu smelters				●		●				●			●

plant for a particular metal, it is likely that the plant is paleoendemic and will not grow except under a very limited set of edaphic (soil) conditions. In only a few cases does the plant actually require the element for growth; most of the time the presence of the element simply decreases competition from other plants, predation, or parasitization by fungi. The high metal concentrations in these plants protect them from both predators and parasites, and the high concentrations in the soil keep competitors away.

The evidence in favor of the slow evolution of the accumulator habit is the almost total absence of accumulator plants in formerly glaciated areas, and their presence in tropical or subtropical areas. The species that have been found to accumulate transition metals to the greatest extent above the soil concentration have been found in the serpentine and copper soils of southcentral Africa and of the Pacific island of New Caledonia. The extensive serpentine area known as the Great Dyke in Zimbabwe is composed of a series of mafic and ultramafic boat-shaped intrusions, and its soils have developed on this ultramafic rock. In these soils, the levels of transition metals of the sixth through eighth groups are very high. The metals found include nickel, iron, manganese, cobalt, and zinc. These areas support a depauperate flora of short grasses, succulents such as aloes, and almost no trees—in contrast with the surrounding normal soils, which are relatively well forested.

Because the African serpentine areas are large, there has been enough room for natural selection to produce truly endemic taxa. In other areas, such as Europe and the eastern U.S., where the contaminated areas are small, the gene pool is too small to produce endemic species, although it can be large enough to produce resistant ecotypes if the time allowed for selection to take place is long enough. The world-wide distribution of serpentine soils suggests that there are many never-glaciated sites around the world that have a high probability of having an endemic serpentine flora, but which have not been explored with these plants in mind. Data obtained from serpentine sites in Cuba suggest that similar endemic species could be found in the other Greater Antilles. Many of these species are likely to be indicators or metallophytes.

Other areas of metalliferous soils are the nonserpentine copper-cobalt belts of southcentral Africa. These are the Shaba province of southeastern Zaïre and the Copperbelt of Zambia. In this area the copper flower *Becium homblei* can be the sole colonizer of high copper soils even though it also grows in nonmineralized soils. The only cobalt-tolerant species known to science are also found in this region, and some of these are hyperaccumulators of this metal. An example of this is *Crassula vaginata* that, although a hyperaccumulator of cobalt, is not confined to copper-cobalt deposits and is found also on the nonmineralized Zaïrean high plateau. Besides this metal-tolerant flora, there are several other nonserpentine metal-tolerant plant communities around the world. One of the best studied is the zinc-tolerant "galmei" flora of Western Europe. Here metallophytes such as *Viola calaminaria* and *Alyssum wulfenianum* have developed in zinc-rich soils, probably from *Viola lutea* and *Alyssum ovirense*, respectively. An extreme example of the galmei flora is a variety of sea-pink (*Armeria maritima*) that, even though morphologically indistinct from the parent species, has evolved to the point of requiring elevated levels of zinc for growth.

Since the metalliferous areas in southcentral Africa are Precambrian in origin, there has been ample time for the heavy-metal-tolerant flora to evolve. On the other hand, extensive glaciations in the Northern Hemisphere have restricted the allowable time for speciation to occur on metalliferous soils to the last 10,000 years, the time elapsed since the last glacial epoch. H. Wild has summarized the differences that should be expected and have been found between the African flora of metalliferous soils and the European one. These are as follows:

(a) The African flora should have many more metal-tolerant species than the European flora.

(b) The African tolerant species should be more distinguishable from nontolerant related species of the same genus.

(c) In Africa there should be a larger number of tolerant populations which have become separate species but remain closely related to the original species.

(d) Some African species should have become so well adapted to the metalliferous soils that they have become completely endemic to them.

(e) In Africa there should be more cases of biotype depletion in which the precursor of an endemic species has disappeared, leaving the metallophyte in an isolated "island."

(f) In Africa, extensive areas of mineralization should have led to the existence of several distinct, but related, endemic species living in different areas.

We have already discussed several examples that highlight some of the points made by Wild. Another example from the European metalliferous flora is the well-known copper flower (*Lychnis alpina*) of Fennoscandia. This plant shows few morphological differences between its serpentinic and nonserpentinic forms. Indeed, the differences in tolerance to heavy metals between the subspecies have had to be established by pot trials. In the African flora, most metal-tolerant taxa have been found growing on soils much more toxic than those found in Europe. Morphological changes are also much more common among African plants.

Survey of Anthropogenic Metalliferous Floras

Contamination of surrounding soils by mining practices differs depending on the minerals mined and the ore in which they are found. In the United States and elsewhere, the soil in and around derelict mines usually contains high levels of the metal previously mined there, as well as of any associated metals leached from the low grade tailings by water and microbial action. For example from table 1 we see that the lead and zinc industries discharge cadmium, copper, lead, and zinc. The copper and nickel industries discharge cobalt, copper, manganese, nickel, lead, and zinc. Smelter-contaminated sites normally have high levels of copper, zinc, lead, cadmium, and nickel. At least some of these metal-contaminated sites have spurred the evolution of metal-tolerant taxa. Others have simply attracted local metallophytes that were already growing in nearby soils naturaly contaminated with these metals. All of these sites have also attracted plants that seem to be intrinsically tolerant of heavy-metal pollution because they possess mechanisms to exclude heavy metals at the level of the roots. Genera that, although not concentrators, are able to withstand high levels of contamination include cattail (*Typha*) and the common reed (*Phragmites*); both genera were found growing in Zimbabwe, in wet regions of a mine dump which contains up to 4000 μg/g of copper. These two examples suggest that it is their wetland habit that has preadapted them to anthropogenic spoils. In chapter 7 we will study this adaptation to wet environments, as well as others.

The rapidity of the evolution of heavy-metal tolerance can be seen in certain grasses. For instance, tolerant strains of colonial bent grass (*Agrostis tenuis*) have developed around refining operations in the city of Liverpool. Before they were more strictly regulated, these refineries contaminated the nearby soil with airborne copper. Some of these soils now contain over 0.2% copper by weight. The oldest contaminated sites are just over 90 years old, and no other plant has yet developed strains resistant to this high level of contamination. Even though biochemical divergence has occurred rapidly under the extreme selection caused by this level of contamination, no morphological divergence appears to have yet taken place, and the metal-tolerant varieties of *A. tenuis* can only be separated from each other or from the parent species by pot trials. This example shows that evolution can take place quickly, but exactly how quickly can only be determined through experimentation.

In one set of experiments, seeds from five populations of creeping bent grass (*Agrostis stolonifera*) growing in sites contaminated by zinc and lead were sown on a copper-contaminated soil, and 5 out of 2000 seeds grew to maturity. These plants proved to be moderately tolerant of copper. For colonial bent grass and an orchard grass, *Dactylis glomerata*, experiments have shown that a few copper-tolerant individuals can germinate from seeds of a nonmine population. These seedlings have tolerance levels comparable to those of populations that normally grow on copper-mine spoils. The high frequency of tolerant individuals in these nontolerant populations suggests that *de novo* mutations are not responsible for that tolerance, but that some individuals of these populations have tolerance-producing genes. Seeds of eight other species that are not commonly found on copper-contaminated soils did not give rise to any tolerant seedlings; evidently, those species do not have tolerance-producing genes. Some species thus seem to have "tolerance genes" at low frequency, at least in some populations. The initial "tolerance gene" might simply be a recessive "defect" in a protein involved in copper uptake. Such a defect might slightly lower the fitness of the plants that are homozygous for this gene when they are growing in normal soil, but would prove extremely useful if they find themselves in copper-contaminated soils. This case demonstrates that an evolutionary advance can be a case of turning a sow's ear into a silk purse. The individuals with these genes are, in a sense, preadapted to at least survive in copper-contaminated soils, and once there, they will embark on a solitary evolutionary path where competition is nil, but where they must constantly combat the poisons at their feet.

An example of an indicator plant that has colonized anthropogenic sites is the grass *Rendlia cupricola*. It has colonized alluvial soils polluted by wastewater from the Likasi treatment plant along the Panda River, in Zaïre. Such anthropogenic sites are also occupied by a cosmopolitan fern, ladder brake (*Pteris vittata*) near Lubumbashi. Another example of the occupation of anthropogenic sites by indicator plants is the interesting case of the African mint *Haumaniastrum katangense*. This mint evolved on naturally occuring copper-cobalt soils, but it has now spread to sites of precolonial smelting activities, and the distribution of the plant has been used to pinpont the location of smelters used by the Kabambian culture in the fourteenth century. This mint has been tried in Texas for removing radioactive cobalt from soil (by continually cropping the plant), but no details have been released.

Because of the slow evolution of the concentrator habit, for plants to become hyperaccumulators takes much longer than for them to become tolerant of metalliferous soils. While hyperaccumulators are extremely rare on previously glaciated areas, tolerant plant taxa are often found there, as well as in other areas of the world. In Europe, river gravels heavily contaminated with zinc and lead are readily colonized by the penny-cress *Thlaspi rotundifolium* ssp. *cepaefolium*. This subspecies also grows on mine tailings and gravel centered on Cave del Predil, a quarry in northern Italy, and a value of 8200 μg/g of lead has been found in plants there. Another plant found in areas polluted with heavy metals is watergrass, *Bulbostylis mucronata* (really a sedge); it is found in temporarily waterlogged soils rich in metals such as zinc, lead, and copper. (The reason many plants can grow in areas contaminated with high levels of lead is the ease with which lead can be immobilized, at the level of the root system, as the insoluble sulfate.) For arsenic-rich gold-mine tailings, the most resistant species are two strains of two different grasses—Bermuda grass (*Cynodon dactylon*) and a related African grass, *Cynodon aethiopicus*—as well as rhodes grass (*Chloris gayana*) and *Leptocarydion vulpiastrum*. For arsenic, a nonendemic accumulator plant is horsetail (*Equisetum*), which has been found to have 781 μg/g dry-weight arsenic when growing over gold-mine tailings in Nova Scotia. (On nonmetalliferous soils, horsetail is not an accumulator.) It is important to note that some species that grow on metalliferous mine dumps do not grow on naturally metalliferous sites. This suggests that metal tolerance is a latent trait in some plants, and that the controlling factors for these plants have more to do with pH or microclimate than with the metal concentrations at a particular site.

In the United States there are several geobotanical indicators of heavy metals. In the eastern and central United States there are indicators of zinc and/or lead. The bladder campion *Silene cucubalus* is an indicator of zinc in New Jersey, and so is mouse-ear cress (*Arabidopsis thalinum*) in Pennsylvania, and sassafras (*Sassafras albidum*) is an indicator of lead in Missouri. In southwestern North America, a gold-poppy, *Eschscholzia mexicana*, is an indicator of copper in Arizona. A derelict copper mine in New Jersey is colonized primarily by herbs and grasses, one of which is a copper-tolerant variety of creeping bent grass (*Agrostis stolonifera*). Soil pH is 4.5 at this site. In a systematic survey of the vegetation growing on derelict gold, silver, copper, lead, and zinc mines in the North Carolina Piedmont, common cattail (*Typha latifolia*) was found to be an efficient excluder of heavy metals, and because of this it could grow in metal-contaminated wetlands. On the other hand the fern *Lygodium palmatum* seemed to be a local metallophyte and accumulated relatively high concentrations of copper, lead, and zinc. Other plants also accumulated heavy metals, but they were no more likely to grow in old mine spoils than in nonmetalliferous areas. For instance, the goldenrod *Solidago pinetorum*, wild carrot (*Daucus carota*), and the empress tree (*Paulownia tomentosa*) all accumulated copper and zinc.

The Lehigh Gap area near the zinc smelter at Palmerton, Pennsylvania is completely barren over approximately 485 hectares and is sparsely vegetated elsewhere. Dominant species are sassafras, black tupelo (*Nyssa sylvatica*), and a sandwort, *Arenaria patula*. A bent grass, *Agrostis perennans*, little bluestem (*Schizachyrium scoparius*), red maple (*Acer rubrum*), and young oak sprouts are also present at this site.

The overall question of how plants attain the hyperaccumulator trait remains unanswered at this time, and there has been little work done on the inheritance of tolerance or of the accumulator trait. There are some species that are innately metal-tolerant, and these species show no differences between the populations found in contaminated soil and those found in noncontaminated soil; an example of this is Bermuda grass.

5 Salty Environments

Introduction

In many deserts, the only semipermanent water is in the form of concentrated brine. It is sometimes so caustic that it stings the mouth and eyes of anyone near it; still, a few organisms can mine the water from it. The plants most adept at mining that water are the chenopods: the members of the goosefoot family (Chenopodiaceae). Their exploits can be easily observed along the shoreline of almost any protected bay. There, two genera of chenopods, *Salicornia* and *Suaeda*, live in areas so salty that a rind of the white substance can cover nearby bare patches of ground, patches too salty for any vascular plant. Some animals, too, are able to live in extremely salty environments, possibly more successfully than plants; brine shrimp (*Artemia salina*) can grow even in the ponds used to extract salt from the sea. Organisms that live in brine have to contend not only with high salt but also with high temperatures and low oxygen. Other such animals include the brine flies so common around the Great Salt Lake; when their adults emerge, their discarded pupal skins make a noxious wrack line around the lake.

As the climate gets progressively drier, brine in the soil turns into salt, and plants must then contend with a lot of salt and very little water. In these salt deserts, the most successful plants are, again, the chenopods; the animals associated with them include not only insects but also tough little rodents. These rodents feed on the salt-encrusted leaves by first scraping off the offending layer with their incisors; this adaptation, as well as others, makes several kinds of rodents common in salt deserts.

Salt Basics

Before we delve further into the ecology of salty environments, it will be helpful to first discuss why salty environments limit plant and animal life. All organisms require a certain amount of sodium, potassium, and chloride ions to establish electrical potentials that allow molecular processes to take place. Most living cells exclude sodium from their cytoplasm and preferentially concentrate potassium, since potassium is much more compatible with enzymatic and other cellular processes than sodium is. An organism living in a concentrated salt solution must maintain the gradients of sodium and potassium by whatever means its genetics allow. The organism may try to keep out the salt by means of some sort of barrier (membrane) that allows only water to pass through. Such selective barriers can be synthetically manufactured for use in desalinization facilities;

therefore, it is reasonable to expect that plants may have come up with the same technique, especially since the material of choice for such membranes is specially prepared cellulose. Barriers are not enough, however. The concentration of salt in the water outside the barrier produces a pull on the water on the other side of that barrier; the higher the salt concentration is, the stronger the pull. This pull has been given the name *water potential*. The water potential is composed of several basic potentials (such as the gravitational potential and the electrical potential) that affect the energy of a water molecule. It can be thought of as the difference in the pull exerted by a solution on a water molecule relative to the pull exerted by a plane surface of pure water on that molecule, and it is given in units of megapascals. What is important is not the actual value of the water potential (which is a relative measure anyway), but the water-potential gradient: the difference between the water potential outside, and the one inside the organism. The water-potential gradient is what concerns us here; because, if the salt concentration outside an organism is large, the water-potential gradient has the effect of drawing water out of the organism and into the outside solution. This pull must be counteracted by the organism, or it will dehydrate and cease to function. The only way to counteract this pull is to produce and concentrate some solute that is compatible with cellular processes. Different families of organisms have come up with different solutions to this problem.

Possibly the chenopods best-adapted to surviving extreme salt concentrations are the glassworts (*Salicornia*). The whole genus is in need of a thorough taxonomic treatment, and now that one of its species has gained a commercial spotlight as a seed-oil crop for desert areas, we may see a thorough revision. Since I am not a taxonomist, I will refer to both the perennial and the annual species as glassworts, and I will simply concentrate on the two lifestyles (annual vs. perennial). These lifestyles lead to completely different tolerances, especially to cold-winter climates. For completeness I will mention that the perennial species are sometimes grouped under the genus *Sarcocornia* by some authors and under the genus *Arthrocnemum* by others.

Preview of Salt-dominated Environments

As we will see in this chapter, there are many areas around the globe that are too salty for most organisms. The salt may come from several different sources, but as its concentration approaches several parts per thousand (ppt), the number of species able to live and prosper in these environments starts to plummet. Why? One factor is the small size of these areas. The smaller an area with a particular set of environmental constraints is, the less chance there is that an organism may have evolved to deal with those constraints. If the salty area is a highly saline pool or a salt lake, another factor acting to limit the number of species adapted to life in it is the variable nature of such bodies of water. Almost by definition, salt lakes precariously teeter on the edge of oblivion, since the causes for their saltiness, high evaporation, and low precipitation can often extinguish a lake for decades or even for centuries. The lake may reemerge in some subsequent wet period, but any life forms that may have evolved in its previous avatar are gone forever. On the other hand, the stability of the oceans has assured a tremendous multitude of species able to withstand salinities from 25 to 40 ppt.

The world's oceans do not impose many physical limits on life, and even high pressure is not intrinsically limiting to life. However, there are a few groups of organisms that have found it difficult to invade the sea. For example, the vascular plants that can claim to be truly marine can be counted on the fingers of one hand. The ocean holds several challenges for vascular plants, and their unique talents, which have made this group of photosynthesizing organisms highly suited for life on land, serve no purpose in the sea. Vascular tissue—a breakthrough in the engineering of water and nutrient delivery in a gaseous atmosphere—is of no use in a water world where each section of a plant can derive what it needs from the surrounding liquid. The elaborate mating structure we call a flower is arguably the masterpiece of land-plant evolution; in it the flagellated sperms' race through a watery course is reduced to its most compact form. But this structure is a clumsy and useless contraption under water. Even with these handicaps, a few vascular plants have managed to invade shallow coastal waters and have become very important in the ecology of soft-bottomed subtidal zones. We will concentrate, as always, on the few that push the limits of endurance.

Polar Mud Flats and Salt Marshes

In the far north, in the realm of sea ice, being a submerged attached organism is not a good idea. Winter scouring will remove most such organisms from all but the most protected embayments. From the arctic slope of Alaska all the way to Greenland, newly scoured mud flats are normally colonized by the alkali grass *Puccinellia phyryganodes*, along with scattered individuals of other species higher up on the shore. In the far north, these companions include several species of sedge including *Carex subspatheca* and *Carex ursina*, and the starwort *Stellaria humifusa*. The only woody species inhabiting these disturbance-prone salt marshes in the far north is the ovalleaf willow (*Salix ovalifolia*).

The only genus of the ones mentioned that has been studied in terms of the physiology of salt tolerance is *Puccinellia* (the alkali grasses). In this genus, a double endodermis provides a barrier to the passage of salt into the xylem. Also, the parenchyma cells in the xylem actively pump sodium out of it and potassium into it, thus providing the leaves with a low internal sodium load. However, there is still the problem of counteracting, with some sort of biocompatible solute, the dehydration caused by the high external salt concentration. To achieve that counteraction, the alkali grasses seem to increase their concentrations of the amino acids asparagine, glutamine, serine, and glycine; they also seem to increase their sugars.

As we travel farther south along the western coast of Alaska, some of the above-mentioned sedges wane, but others take their place. A few more grasses such as *Dupontia fisheri* and the reed grass *Calamagrostis deschampsioides* join the community, but none of these displaces *P. phyryganodes* as the primary colonizer of the mud flats. By the time we reach the Seward Peninsula, Lyngby's sedge (*Carex lyngbyei*), has become a common member of the climax community above mean sea level. By this latitude, a few more woody species, including *Potentilla egedii*, have begun to infiltrate the mud flats and salt marshes. *Potentilla egedii* can be found as a colonizer on the bare mud, or more commonly, higher up on the shore in a zone that is becoming dominated by several species of plantain (*Plantago* spp.).

Unlike the alkali grasses, the plantains use sorbitol as their biocompatible solute. To protect the plant from the increasing salt concentration in the leaves, the concentration of sorbitol increases in step with that of sodium chloride (the sodium chloride is sequestered in the vacuoles of the cells). Sorbitol has at least one drawback as a biocompatible solute even at low concentrations, it sharply increases the viscosity of water solutions. Whether this viscosity increase limits the possible northern distribution of the plantains is not known, but the arctic members of the genus do not occur in saline environments.

Farther south along the gulf of Alaska, the main mud-flat colonizer is still the alkali grass *P. phyryganodes,* which forms open mats on the mud flats. But other colonizers have begun to become commonplace; among these are two cosmopolitan annual chenopods: the glasswort *Salicornia europaea*, and the sea-blite (*Suaeda calceoliformis*).

The chenopods are well-represented in almost all saline environments. However, their particular physiological approach to high external salt, namely, their concentrating the salt that leaks into the root and excreting it through leaf pores, seems to have left them vulnerable to frost damage. It is revealing that the northernmost station for a perennial member of the *Salicornia* clan is in the Alaskan panhandle.

Higher up on the shore, above the mud flats, we find a second zone. Tufts of another alkali grass (*P. trifolia*) appear, along with scattered specimens of the sea-milkwort (*Glaux maritima*) and the Canadian sand-spurrey (*Spergularia canadensis*). This second zone is dominated by several species of plantain, and also by the seaside arrow-grass (*Triglochin maritima*), which has started to insinuate itself among the sedges. (It will be discussed at length as part of inland salt-controlled ecosystems.) This zone can be very wide and bare, suggesting that in it, the smaller influence of tidal flushing concentrates salts and prevents aeration of the sediments. It can be dominated by very few species, including *Plantago maritima*, *Stellaria humifusa*, and *Carex lyngbyei*. For example, in places like Kodiak Island, the plantains, together with *Stellaria humifusa* and *Carex lyngbyei*, constitute 92% of the plant cover. The presence of *Salicornia europaea* in low, poorly drained spots indicates high salinity and scarce oxygen. In those spots, microtopography becomes extremely important, and an elevation of a few centimeters can mean the difference between having a mound of lymegrass (*Elymus arenarius*) or having a muddy hole covered with a carpet of glasswort.

A third zone develops where the influence of salt is lessened by the height above sea level and by freshwater influences. It is dominated by the seaside arrow-grass and by *Potentilla egedii*. This zone develops on less-brackish and better-drained soils, and it sometimes coincides with the driftwood zone seen on many Pacific beaches south to Washington and Oregon.

Continental Salt-controlled Ecosystems

We will now head inland to those places inside the North American continent that, for one reason or another, have accumulated salt over thousands of years. These places stand as lonely outposts of several species whose nearest relative might be thousands of kilometers away. As we move inland from Yakutat on the Gulf of Alaska we pass over the St. Elias Mountains which ring the moisture from the warmer air coming in from the Gulf of Alaska. As we descend into the drier valleys on the other side of the range, we first encounter Kluane National Park and the delta of the Slims river as it spreads out into Kluane Lake. This vegetated delta is composed of fine silt that during the hot dry summers is pockmarked by shallow pools drying in the sun. It is an alkaline grassland dominated by Nuttall's alkali grass (*Puccinellia nuttalliana*) and *Plantago maritima*. The alkali grass is far to the west of its closest population on the Canadian arctic shore.

Another inhabitant of these warmer and drier valleys is the northernmost race of winterfat (*Ceratoides lanata*). (If we were to follow proper rules of precedence, the name of this genus would be *Krascheninnikovia* after an eighteenth-century Russian botanist, but I will retain the *Ceratoides* for the sake of familiarity.) We will meet another *Ceratoides* in the limiting environments that are the dry and lofty high central Asian plateaus. *Ceratoides* is another chenopod; that is, a member of one of the preeminent plant families of relatively cold and salty deserts. In their evolution in warmer climes, this family confronted, and adapted to, soils high in gypsum as well as in other salts. Its adaptations included internalizing sodium chloride to reach osmotic balance with the outside environment. We will meet many more chenopods in this book, but here only a lucky coincidence is of interest: by internalizing salt and increasing the osmotic pressure inside cells, a plant cannot only deal with salty soils but also deal with cold, up to a point.

No matter what the nature of a substance is, dissolving it in water will decrease the freezing point. This freezing-point depression, as it is called, depends only on the ratio of water molecules to solute particles, and it is a colligative property of the solution. Chenopods have adapted to salty soils by sequestering elevated levels of sodium chloride somewhere in their cells. Most likely, the sodium chloride is sequestered in vacuoles, and the osmotic pressure of the cytoplasm is adjusted to match that of the vacuole by adding biocompatible solutes. This expedient has preadapted chenopods to cold environments in which the cold is not too intense. A 23% table-salt solution will have a freezing point of -20.5°C, and this solution is saturated at its freezing point. This sets a theoretical limit on the ancestral preadaptation of chenopods (especially American atriplexes) to freezing temperatures. This limit is merely academic, since no higher plant can withstand salt concentrations that even approach a value of 23%. Unlike glycerol, sodium chloride is not very effective at producing supercooling (see chapters 8 and 9), and it is not a cryoprotectant. The high internal sodium chloride concentration poses a risk to these plants in winter. Salt becomes less soluble as the temperature decreases. At low temperatures, the presence of precipitated salt crystals can block the slow flow of water and of oxygen that maintain perennial plants through winter. If the temperature drops low enough so that the sodium chloride in the plants' cells precipitates, the cells will most likely be damaged. Chenopods living in these areas must use other mechanisms besides freezing-point depression, but only a few of them seem to have evolved that far in their post-glacial march north. For instance, the only members of this family that reach the American arctic are annuals, which avoid the winter cold by going to seed. The perennial members of the family especially the perennials belonging to the genus *Atriplex* are not far behind, however. In Canada, one perennial atriplex, Nuttall's atriplex (*Atriplex nuttallii*), has made it down the Peace River drainage to the heart of the boreal forest at Wood Buffalo National Park. Certainly, to be able to withstand the -40°C of its northern range, Nuttall's atriplex must have evolved a more effective freezing-tolerance mechanism than that conferred by its ancestral salt tolerance. For these halophytes, moving northward from their present location will be difficult, since low precipitation and concomitant salt buildup in the soil are rare in the moist woodlands to the north. Even at Wood Buffalo, the salt lakes and flats are possible only because of a large natural salt deposit, a relic of an ancient desert. However, with the propensity of atriplex plants to hybridize and create new opportunities for themselves, it is likely that this genus will keep on marching north. As we will see, chenopods and graminoids (a term for a convenient grouping of grasses and grasslike plants) are the preeminent salt-adapted plants of middle latitudes.

The Salt Plains at Wood Buffalo

As we pass over the northern inland valleys of British Columbia and meet the northern Rockies, we encounter the Peace River, born of them. Following this tamed river westward through farmland and into the white-spruce floodplain forests, we come to the most diverse inland deltas in North America. Almost as an afterthought, the Peace River starts heading southeast at Boyer Rapids. This fateful turn takes the river on a collision course with the Athabasca River coming from the south. Their collision and interdigitation form a vast landscape of ponds and marshes; the area is a herald of Lake Athabasca, which opens just east of the confluence. This inland delta is the last wild breeding place for the whooping crane (*Grus americana*). It is also the summer home for 230 other bird species. Our interest lies a little to the north of the delta, along the flat expanse known as the Salt Plains. This 370-square-kilometer area (dotted with salt cones one-meter high and salt pans hectares in extent) is the product of millions of years of geologic activity. Two hundred and seventy million years ago, the area just southeast of the Salt Plains was part of a vast, shallow inland sea. As the sea dried, the salt was concentrated and eventually buried. For millions of years rain has fallen on the Caribou Mountains, penetrating the ground and traveling northwest following the slope of the land, eventually encountering the salt deposits. The salt-laden ground water continues northwest until it meets the Canadian Shield. The flows then bubble up to the surface as springs, unsure of the way to the sea. Springs braid through the flat expanse of the Salt Plains until one by one they are corralled by the local topography and collected into the Salt River. This river lives for only about 30 kilometers before it itself is captured and its salt riches stolen by the Slave River heading north.

In spring, both sexes of the long-nose sucker *(Catostomus catostomus)* don bright pink-red hues and prepare to spawn by migrating a short distance up the Salt River. They are taking advantage of the refuge from egg predation provided by the higher salinity of its waters. Their spawning in its waters and then returning to the fresh waters of the Slave River to feed gives them a convenient nursery.

The vegetation of the salt flats within the Salt Plains is poor and consists of species we have encountered on the coast, including the annual chenopods *Salicornia europaea* and *Suaeda calceoliformis*. In the fall, the *Salicornia* plants on these flats seem to turn an even deeper shade of scarlet than that of the plants on the coast, giving the Salt Plains their characteristic autumn hue. That is one of the reasons why some authors segregate this glasswort into a different species, *Salicornia rubra*.

Salt Lakes

Salt lakes occur all over the world wherever high evaporation, low precipitation, and localized dead-end (endorheic) drainages coincide. There are several salt lakes that have been extensively studied, possibly because of their sheer inhospitableness; among them are the Antarctic salt lakes (they'll be covered in chapter 8). The Arctic also has salt lakes, but they have not been studied much, and little information is available about them.

Canadian Salt Lakes

Although highly saline lakes are not common north of the 50th parallel, south of this latitude evaporation can exceed precipitation; this happens in several Canadian provinces. Salt lakes may be found in Alberta, Saskatchewan, and British Columbia.

The Canadian Prairie Provinces (Alberta, Saskatchewan, and Manitoba) are subject to more evaporation than precipitation (35-cm precipitation vs. 100-cm evaporation at the lakes we will discuss), and the many endorheic drainages lead to saline lakes. (A few of these lakes are almost as salty as Utah's Great Salt Lake.) Most of these lakes are covered by a meter thick sheet of ice from mid-November to May. Several researchers have studied over 40 saline lakes in the southern parts of Alberta and Saskatchewan in order to find a series of similar lakes of increasing salinity. Such a series of lakes should show, in the clearest fashion, the limits that salinity imposes on organisms. You might wonder why this approach might be better than a laboratory approach where organisms are exposed to different salinities, and a determination is made of their salinity tolerance. The answer is that laboratory investigations are necessary, but the nuances of the real environments are such that an organism that may withstand a certain salinity range in the lab may be restricted to a much smaller range in nature. The opposite may sometimes happen, but it is a much rarer occurrence.

In southern Saskatchewan, some of the saline lakes are soda lakes, which have a unique assemblage of plants and animals, but these lakes have been studied little. Other saline lakes in this area are managed for fishing or as stopover sites for migrating waterfowl. Before this type of management was begun, the salinity in some of these lakes fluctuated wildly because of the yearly rhythms of precipitation and evaporation. Now some of these lakes are kept at a constant water level, and this translates into a constant (and relatively low) salinity.

There are several reasons why organisms may have been enticed into progressively saltier waters; two of them are to escape predation and to escape parasitism. An example of the former has been discussed in relation to the long-nose sucker. Let us now look at an example of the latter. In lakes of intermediate salinity in British Columbia, two water boatmen of the genus *Cenocorixa* share a mite parasite. *Cenocorixa bifida* is fairly tolerant of the mite parasite, but *Cenocorixa expleta* suffers heavy mortality when parasitized. Therefore, *C. expleta* only thrives in saltier waters where the mite cannot survive. This may be a case where a parasite has made its host evolve a higher salt tolerance because of extreme selection pressures.

In the most saline lakes in southern Saskatchewan, such as Lake Chaplin (169-222 g/l) and Lake Muskiki (232-370 g/l), we find a very simple planktonic ecosystem. Both lakes have the green alga *Dunaliella salina*. Lake Chaplin has a filamentous green alga, *Ctenocladus circinnatus* (Trentepohliales), which sometimes occurs in great abundance, and it also has rotifers and brine shrimp. (A similar planktonic ecosystem is found in Utah's Great Salt Lake.) The salinity of Lake Muskiki is too great even for brine shrimp, and in it only the green alga *Dunaliella salina*, thrives. Slightly less saline lakes such as Little Manitou (55-121 g/l), which has been a health spa for many decades, and Aroma (97 g/l), have a more varied fauna. Besides having brine shrimp, they also have benthic animals, including several species of brine flies (*Ephydra* spp.) and long-legged flies (dolichopodids). Along the shore, the water scavenger beetle *Enochrus diffusus*, feeds on the algae in the water, as sometimes do other insects such as the water boatman *Trichocorixa verticalis interiores*. (This water boatman is a subspecies of the common water boatman *Trichocorixa verticalis*, which has been collected from Delaware Bay to the Gulf of Mexico.) It seems likely that 120 g/l is close to the limit of salt endurance for most of these insects, at least in the sodium-sulphate-laden waters.

None of these highly saline lakes has submerged hydrophytes of any kind. When the water level fluctuates considerably, two species of bulrushes (*Scirpus paludosus* and *Scirpus americanus*) venture into these deadly waters. The fluctuations in the water level are due to precipitation, which forms a layer of brackish water on top of the brine. When it rains, the bulrushes are inundated. They are exposed to diluted lake water and can grow during that time and then remain inactive during the drier, saltier periods. The other vascular plant species that withstands the harshness of salt-encrusted soil around these lakes, even without the flushing effect of large fluctuations in water level, is the annual glasswort *Salicornia europaea*.

Higher up the shore, the dominant plants change, giving the landscape contours of different colors and textures. For example, the glasswort starts giving way to the seaside arrow-grass (*Triglochin maritima*), the poisonous perennial that we encountered on the northern Pacific coasts of North America. This plant's leaves make it look like a rush, but its flower spike shows that it belongs to its own family, the Juncaginaceae. The seaside arrow-grass prefers soils coarser than those the glasswort prefers, maybe because coarser soils allow it to produce deep roots that will eventually reach the water table. Then, the plant will not be at the mercy of the feeble and inconstant surface-water supply that may evaporate by the middle of summer leaving only a salt crust.

Depressions found in coarser soils in southern Saskatchewan may have very salty soil at their centers, and because they do not hold water, these centers may be completely devoid of vascular plant life. The glassworts that live in these depressions could invade the centers, were it not for the lack of surface-soil moisture, which inhibits the germination of their seeds. These depressions are in effect the site of a confluence of limits that no vascular plant in the area can overcome. Most of the halophytes in the area are limited by the high salt content of the centers of the depressions. The few halophytes that can grow in the depressions are annuals that must reestablish their presence every year. If the center of a depression is not conducive to germination, the depression will remain bare in all but the wettest years. The predominance of annual halophytes in this area, which is deep within the North American continent, suggests that the physiological mechanism these plants use to balance the high external-salt concentrations are incompatible with cold hardiness. The halophytic members of the goosefoot family found in the northernmost salt meadows (well-vegetated salt flats) in North America are *Salicornia europaea*, *Suaeda calceoliformis*, *Chenopodium rubrum*, *Chenopodium salinum*, and *Atriplex hastata*. These species are all annuals, though farther south, in the salt-desert areas of the Great Basin, each of these four genera has perennial members. All halophilic chenopods seem to use the previously mentioned uptake-and-secretion mechanism to compensate for the low water potentials of the soil solutions in which they grow. This mechanism makes many of these chenopods ready sources of table salt. *Salicornia* in particular has been used since ancient times in Europe and Asia as a source of soda for glassmaking, hence the name *glasswort*. Through a combination of high salt, lack of water, and cold winters, the depressions where the glasswort grows become an extreme environment near the center of the North American continent.

The periphery of the depressions discussed above is reasonably well vegetated (5 to 17%) by a monotypic stand of the seaside arrow-grass. This deep-rooted perennial need not worry about the availability of surface moisture for growth. It, however, must cope with the soil's high salt concentration, and it copes by producing and hoarding proline as its biocompatible solute. Proline, an amino acid, is used as an osmotically active solute by several other unrelated plant families. It is also quite a bit more soluble in water than sodium chloride; therefore, it diminishes any risk of precipitation as the temperature drops. The members of the arrow-grass family have other interesting chemical adaptations besides the use of proline. In case of drought, or other times of stress, these plants produce high quantities of cyanuric glycosides, presumably as grazing inhibitors. As we will see in chapter 7, cyanuric glycosides are not compatible with a wetland existence, but this plant apparently produces them only under relatively dry conditions. These plants are not always poisonous, however, and once they are cut, hay made from them loses hydrocyanic acid rapidly upon drying.

In the saline valleys of northeastern California, a different species of arrow-grass, *Triglochin concinna*, is often mowed for hay to feed cattle. The cut stems and leaves are left in the fields to dry, in order to reduce their cyanide content. An instance of how deadly these plants can be was recorded in these hay fields in 1959, when many jackrabbits died after eating the freshly mowed plants. All the dead animals were found within several hundred meters of the field.

United States' Salt Lakes

Utah's Great Salt Lake is a well-studied hypersaline basin that has a picturesque and geologically significant past. What we now know as the Great Salt Lake is the small remnant of the great Lake Bonneville of late Pleistocene times. At its most extensive, this lake covered most of what is now the western half of Utah north almost to the Snake River. It was a shallow brackish lake that must have supported many species of fish as well as submerged and emergent vegetation. By the time Native Americans encountered the lake sometime in the last 10,000 years, it had retreated from most of the land it had once dominated and probably was about as extensive as it is today. The pace of change in the lake accelerated with the building of a railroad causeway in the nineteenth century. This causeway in effect created two different lakes: the northern basin is highly affected by freshwater influx from the mountains, and it is therefore much fresher than the southern basin which has meager freshwater influences. The salinity of the southern basin (which averages 20%) precludes all but the simplest of ecosystems. The main plant in it is the green alga *Dunaliella viridis*, and the only zooplankter is the brine shrimp. *Artemia salina* is one of the few fairy shrimp (family Chirocephalidae, order Anostraca) that can stand salinities as high as that of seawater or higher. It is a cosmopolitan fairy shrimp that thrives in hypersaline waters throughout the temperate and tropical regions of the globe (with the exception of Australia, where a related species lives). Under the right conditions, it may be found in the hundreds of thousands of individuals per cubic meter of water. What allows this defenseless creature to reach such densities is its choice of habitat and its method of reproduction.

Brine shrimp can grow in waters that have one tenth the salinity of seawater, but so can many of its potential predators. It is a slow swimmer and has no defenses against predators, a trait that makes it unable to compete in waters of low salinity. As the salinity increases past that of seawater, the number of species able to survive for prolonged periods diminishes almost exponentially. However, coastal ponds fed by storm tides can contain brine shrimp. The physiological adaptability of brine shrimp makes this species uniquely suited to fluctuating salt concentrations. In the summer at tropical and subtropical latitudes, a shallow coastal pond may start out with the salinity of seawater, but that salinity will soon double or triple. Most other aquatic creatures cannot adjust to such variations in salinity and soon die.

There are two ways for an organism to survive wide salinity fluctuations. An organism may regulate its internal salt concentration at a certain level, regardless of the external salt concentration. Such an organism is known as *an osmoregulator*, and it uses active tissues and organs to pump the intruding salt back out into the environment. On the other hand, an organism whose internal solute concentration varies in response to the external concentration is *an osmoconformer*. The brine shrimp is an osmoregulator up to a certain external salt concentration; then it can turn into an osmoconformer. Every aquatic organism can physiologically deal with a particular salinity range; outside of this range, such an organism will die. The brine shrimp, along with the brine flies of the genus *Ephydra*, can withstand extremely wide salinity fluctuations, and in an ecological sense depend on them. Both these creatures feed on the algae that also thrive in hypersaline waters. *Artemia* occurs in such habitats all over the world, but it is sensitive to temperature extremes. In the more permanent salt lakes of the world, the populations of brine shrimp have evolved unique adaptations to the local conditions. For instance, the brine shrimp of Mono Lake in California are unique in that they produce dense overwintering eggs that sink to the bottom of the lake at the end of the active season, whereas the brine shrimp of the Great Salt Lake produce lighter overwintering eggs.

During the spring, the increasing temperature and light intensity cause the algae to bloom; and the brine shrimp, with more algal food available, explode in a profusion of mating and egg-laying. These spring and summer eggs are quite different from the tough eggs of the previous fall. In the Great Salt Lake, brine shrimp lay nondurable eggs during their active growth phase, which lasts through the summer. As in many simple ecosystems, the population explosion of the shrimp ends as the food supply is exhausted, and the population declines rapidly in late summer. By fall the few shrimp that are left, possibly cued by the shortening day length, produce overwintering eggs. The adults and larvae then die by mid-October, as the surface waters cool down to between 6 and 9°C, the lethal limit of this fairy shrimp. Though the algae are now free to reestablish their numbers, the temperature and low light levels only allow a minor increase in algal numbers. They must wait until the following spring to again race ahead of the shrimp, only to be overtaken by midsummer.

The unique ability of the brine shrimp to live in hypersaline waters can be attributed to its ability to produce glycerol along with other compatible solutes; these solutes allow it to equalize the water potentials inside and outside its body. Since brine shrimp can live in hypersaline waters close to the saturation point of sodium chloride, the concentration of glycerol they must produce is substantial. To live in the Great Salt Lake, they need to produce 8 to 10% glycerol by weight. This concentration of glycerol does not help the brine shrimp survive cold temperatures; this provides an example of how poorly we understand the role of glycerol as an osmotic solute and as a cryoprotectant.

In slightly less salty water, other animals join the brine shrimp in the algal feast. This happens in the Great Salt Lake, where two species of brine fly, *Ephydra cinerea* and *Ephydra hians*, scrape the algae off the rocks in the shallows. These two fly species are also present in Mono Lake even though it is a soda lake (the Great Salt Lake has salts in nearly the same proportion as seawater, i.e. it is nearly thalassohaline).

In soda lakes, the concentration of carbonate and bicarbonate is much higher than in other saline inland waters. These ions cause a high pH that is inimical to life, and even the brine shrimp that inhabit Mono Lake have had to evolve to deal with this toxic liquid. An example of that inimicalness involves populations of the brine fly *Ephydra hians* in Mono Lake, California and Lake Abert, Oregon. The Pacific coast of the United States received unusually high precipitation in the winter of 1983-84. The following spring, Mono Lake and Lake Abert experienced a decrease in their salinities; the former dropped from 90 to 80 g/l, and the latter dropped from 30 to 20 g/l. The increase in brine-fly population in Mono Lake after this dilution may mean that the population was salinity-limited previously, and the decrease in salinity allowed it to increase its numbers, but the salinity was still too high for most other organisms. On the other hand, Lake Abert's decrease in salinity ameliorated the stress for other organisms and allowed them to prey on, and compete with, the brine-fly.

Salt Creek and Cottonball Marsh

It has often been said that Death Valley is one of the most inhospitable places on Earth. This certainly seems true when one steps out of an air-conditioned vehicle and experiences the oven-like heat of summer firsthand, but Death Valley also has its wet side. There are several marshes in the valley, the largest of which is Cottonball Marsh. It lies at an elevation of 80 meters below sea level and is composed of two separate areas, together encompassing less than 300 hectares. The water for the marsh comes from groundwater seeping up through nearby faults. In winter, even such a steady source manages to cover only 40% of the marsh with a few centimeters of water, except for some pools that may be up to a meter deep. In summer, evaporation is much higher, and standing water shrinks down to 10% of the total marsh area. It might be said that to call this area a marsh is a stretch of the imagination, yet marsh organisms do survive here. The vegetation consists of a few clumps of iodine bush (*Allenrolfea occidentalis*) and salt grass (*Distichis spicata*) growing close to the water-seepage areas, where the salinity is lowest. Farther down the alluvial fan, away from the seepage, the ground is covered with a crust of calcium and sodium sulfates. The least saline of the semipermanent pools harbor a few blades of ditch-grass (*Ruppia maritima*) and Cooper's rush (*Juncus cooperi*). Along the edges of the pools, algae precipitate gypsum and form small domes under which the amphipod *Hyalella azteca* and the snail *Tryonia* graze. Ostracods—minute ancient crustaceans with a hinged shell—like tiny animated clams scurry about at breakneck speed.

Pupfish (*Cyprinodon milleri*) too are able to live in these pools. These little fish (each only 3.8 cm long) belong to the same genus that inhabits nearby Salt Creek, and are in fact a splinter population isolated in Cottonball Marsh for possibly up to a few thousand years. In that time this population has acquired several distinctive characteristics, not the least of which is its ability to withstand salinities equivalent to 110 ppt NaCl. This is approximately equivalent to the maximum salinities withstood by the insects in the Canadian lakes.

Little is known of the breeding behavior of the Cottonball Marsh pupfish population, but the Salt Creek population has been studied. During the breeding season, the male has deep blue sides and an iridescent purple back. The fish breed whenever conditions allow, and with a generation time of only 2 to 3 months, the population in Salt Creek may reach into the hundreds of thousands. This is an amazing density when one considers that the pupfish occupy only a 2.4-km stretch of the creek. (The marsh population presumably does not reach such levels.) These little fish may live only ten to twelve months; therefore, the cycling of nutrients in the pools they inhabit is quite fast. The fish eat the ostracods and amphipods that eventually eat the fish.

Flamingos

In some tropical regions, including the Andean highlands and the African rift valleys, the salt lake ecosystems contain remarkable filter feeders that, like some miniature whale, sift the brine in order to catch the organisms suspended in it. Some of these creatures are the top predators of these ecosystems, feeding on the brine flies and brine shrimp, while closely related species are primary consummers like the brine shrimp themselves. I am speaking of those curious, yet charming birds—the flamingos.

There are three genera of flamingos, but they all have in common certain characteristic traits including a bill unique among birds in the extent to which it has been adapted for the job of filter feeding. The upper bill of flamingos is curved. In the species that feed on the bottom muck, such as the greater flamingo (*Phoenicopterus*), the bill fits like a lid over the cavity in the lower bill. Flamingos with this bill design sift the bottom muck for arthropods and other small animals including the killifish (*Cyprinodon* spp.) that sometimes make briny coastal pools their home. The bill and tongue of these birds operates like a piston pump. As the tongue is pulled back, water rushes in, pushing down hair-like laminae as it flows in. In the reciprocal motion, the tongue is thrust out, sending water rushing out, but this time the laminae stand up and block the escape of small organisms. Any particle greater than the separation of the laminae, but smaller than a few centimeters, is fair game.

On the other hand, the lesser flamingos (*Phoeniconaias*) have evolved to feed on the algae that provide sustenance to the brine shrimp and brine flies themselves. The bill of the lesser flamingo does not possess a large cavity and the upper bill fits quite snugly into the lower one. The laminae are also much more closely spaced than in the greater flamingo. Whereas the greater flamingo may have 12 to 15 laminae per centimeter, the lesser flamingo will have from 45 to 50. These birds are only found in alkaline lakes in Africa, the Middle East, and India. In the eastern rift valley lakes of Africa, their diet consists mostly of *Spirulina*, with small amounts of *Dunaliella* and diatoms. They feed by skimming off the surface, never dipping their bill more than a few centimeters. Using this method they can feed while swimming in water that is too deep for them to stand in.

The greater flamingos are limited in this regard, and must be content with walking and feeding in relatively shallow water. These birds are more coastal than the lesser flamingos, and the American populations frequent brackish lagoons along Caribbean shores. They have a spotty distribution and are found in the Galápagos, the peninsula of Yucatán, the northern South American coast, and the Bahamas. Members of the Bahamian population are occasionally sighted in the Greater Antilles and even in Florida.

Australian Salt Lakes

Each continent's saline lakes have their own style, their own definite pattern of life; those in Australia are no exception. In them, the ecological niches occupied by ephydrids, water boatmen and flamingos elsewhere are occupied by other types of organisms. Australian saline lakes contain a preponderance of flies including crane flies, biting midges, and chironomid midges such as *Tanytarsus barbitaris*; they also have predaceous diving beetles including *Rhantus pulverosus* and *Lancetes lanceolatus*.

The brine shrimp is absent, but another genus of fairy shrimp, *Parartemia*, has diversified to take up the slack. There are eight recognized species in this genus, all from Australia. They have been found in lakes with salinities up to eight and a half times that of seawater.

Another Australian member of the salt lake fauna is the salt lake slater (*Haloniscus searlei*). This relative of woodlice (terrestrial isopods) took to the brine in the distant past and now can be found in lakes with salinities up to five and a half times that of seawater, and in waters up to 14 meters deep. Isopods are very rarely found as members of salt lake communities in other parts of the world, and this is just one example of the uniqueness of Australian salt lakes. Unlike many other inhabitants of salt lakes, it passes the dry season as an adult, hiding in damp microsites while it waits for the rains to return.

Other organisms that survive the dry season as adults are the salt lake snails in the genus *Coxiella*. Like the slaters, these snails survive the dry season by employing avoidance strategies; in their case, they close their shells and aestivate in a cool microsite.

One particular area of Australia that has been relatively well studied in terms of its salt lakes is the Eyre Peninsula. In this area we find some of the animals we have already mentioned plus a couple of surprises. One of the most intriguing of these is a salt lake spider that can be found on submerged vegetation at a depth of 25 cm, some distance from shore. It appears to use an air bubble to breathe.

Microbial Halophiles

The organisms best adapted to a life in saturated solutions of salt are the halophilic purple bacteria. These unusual bacteria do not seem to be closely related to any other group, although over the years different authorities have claimed different affinities for them. For starters, their plasma membrane is not made of lipid esters like that of most other bacteria but of *hydrocarbon ethers*. (There are a few other bacteria that shun lipids; they live in acid hot springs throughout the world.) These bacteria also possess a unique form of light-to-energy conversion that is unlike any of the forms of photosynthesis found in nature and more similar to the mechanism behind vision in vertebrates. The cellular machinery responsible for this light-to-energy conversion is produced when these bacteria find themselves in a low-oxygen environment (as happens often in saturated brine). Briefly, these bacteria's mechanism uses a purple molecule that is a pigment-protein-membrane complex similar to rhodopsin (the visual pigment of most vertebrates) to pump protons from one side of the cell membrane to the other. (This is akin to what happens in a photovoltaic cell as electrons and holes are "pumped" across a barrier by each photon of light.) Once this proton gradient is established, there needs to be a way that it can be tapped. Again, an analogy would be the photovoltaic cell, where if the circuit is open, no benefit is derived from the device. On the bacterial cell membrane, special proteins form pores that allow a controllable flow of protons back into the cell. These membrane proteins act like the turbines in a hydroelectric plant; every few protons that come back into the cell generate a little bit of usable energy in the form of ATP. This ATP is the energy currency of all free-living organisms.

These purple bacteria are obligate halophiles, and they require salt concentrations approaching saturation. They can be found living in the seawater evaporation basins used to produce salt from seawater, in conditions so severe that only microorganisms can survive. Once the concentration increases beyond 23% salt, there are no brine shrimp or brine flies, and the shores are thick crusts of white salt. The purple bacteria must reach osmotic equilibrium with their surroundings; otherwise they would shrink into microscopic prunes in the mummifying fluid of those salt-saturated ponds. Unlike eukaryotic organisms, these bacteria use potassium chloride as their compatible solute, and, to reach osmotic equilibrium, they have to concentrate potassium several hundredfold over the external concentration.

The purple halobacteria are not the only light-harvesting organisms in those salt-saturated ponds. We have already encountered two species of *Dunaliella*, a genus of green alga belonging to the family Volvocales, which also makes its home in saturated sodium chloride. Unlike the purple bacteria, these algae are eukaryotes; they have a nucleus and other organelles that make them more closely related to us than to the bacteria with which they share the ponds.

Dunaliella's mechanism for equalizing the osmotic potential is completely different from that of the purple bacteria; it uses our old friend glycerol to accomplish this feat. The concentrations of glycerol that it must use are so high—around six molal—that it surpasses anything we will discuss regarding insect winter tolerance. (At these glycerol concentrations, the algal cytosol's freezing point is depressed to -20.5°C or below, so their cold susceptibility is possibly due to a membrane effect.) In many ponds, this alga is the dominant organism in the phase right before saturation, whereas after salt starts to precipitate dominance is usurped by the purple bacteria.

There are other single-celled eukaryotes that can withstand even lower water potentials in the external medium. These include several yeasts related to brewer's yeast. *Saccharomyces rouxii* is an osmophilic yeast that makes its home in the honey of bees or in the confections of humans. The water potentials reached by those substances are so low that, were they made with an ideal solute, their freezing point would be close to -70°C. To live in those substances, yeasts and fungi must have extremely low internal water potentials; because of the special solutes they use, their internal solutions will not freeze until very low temperatures are reached. For this reason it is not surprising that prokaryotes, fungi, and yeasts are the only living organisms to be found in Don Juan Pond, a hyperhaline Antarctic pond, whose $CaCl_2$ concentration is so high that the water does not freeze until -48°C.

Coastal Lagoons

A lagoon can be defined as a shallow body of water connected to the ocean through a channel that restricts water movement. Lagoons may contain many communities normally associated with estuaries, except seagrasses, many of which need stronger currents than those available in lagoons.

Lagoons can often be extremely productive places. The two most important factors behind this high productivity are the following: (1) A lagoon's shallowness (normally less than 10 m at its deepest) allows the photic zone to encompass the whole depth and allows the wind to mix the water column (in most circumstances). (2) A lagoon's restricted access dampens out fluctuations in the salinity profile and dampens out the tidal range. These factors give a lagoon a stability that allows producers to take up residence in environmentally optimal sites and not be stressed by drastic diurnal changes. On the other hand, these same factors make parts of a lagoon extremely challenging environments. These differing aspects of a lagoonal ecosystem are tightly intertwined, and they serve as a small-scale example of how extreme environments are intricate parts of whole ecosystems.

The main source of food for the lagoon ecosystem is the detritus pool, which can be derived from several sources. Most lagoonal animals consume detritus, benthic algae, and mangrove root epiphytes indiscriminately. This is the case for economically important species such as shrimp, which spend part of their life cycle in these lagoons and migrate out of them when they are from half to full grown. The native Hawaiians practiced fish aquaculture in natural and man-made lagoons which they fertilized with cut grass and organic marine debris, thus increasing the detritus pool. They apparently could achieve yields of 400 kg/ha by this method.

A key to a lagoon's productivity is the way it handles phosphorus. Lagoons, such as the Caimanero Lagoon in Mexico, act as phosphorus traps. This lagoon exports most of its phosphorus in the bodies of the organisms that spend their juvenile stages in it (fish, shrimp, etc.). Apparently, only about 8.5% of the postlarvae that enter the lagoon are harvested by fishermen when they come out. In lagoons, a major method of internal recycling seems to involve the growth of halophytes, especially the perennial glasswort (*Salicornia subterminalis*) and the saltwort (*Batis maritima*), during the dry season when the water level drops. These plants then incorporate phosphorus from the sediments and release it (they release 500 mg P/g dry wt.) by decomposition when the desiccated areas are flooded.

A likely explanation for the very large precipitation-phosphorus input into many tropical lagoons near agricultural areas is the deposition by rain of particulate phosphate derived from fertilizer use. Such nutrient addition can change the dominance of the primary producers. For example, in one case of fertilizer input, the red algae known as tubed weeds (*Polysiphonia* spp.), increased from trace to 15% coverage. Some of the *Polysiphonia* escape the lagoons and end up on the menu of reef fishes like the surgeonfish (a common coral-reef fish). The surgeonfish in turn are considered delicacies in the Indo-Pacific region.

In areas where evaporation is greater than precipitation (such as the southern Mediterranean), the tenuous connection with the sea can be lost if the outflow of water is not sufficient to keep the channel open against the ever-present movement of sand, thus making the lagoon a closed-evaporation basin. Constant human interference (dredging of the outflow channel) is the only thing that prevents this fate. This measure is enough to keep some very productive fisheries flourishing.

There is a mutually beneficial interaction between humans and productive lagoons in many places on the planet. This is especially true where fisheries are less energy-intensive, and therefore fish are caught close to shore. Even in developed countries, substantial reduction of lagoonal area would hurt the national economies and the local economies by decreasing the catch-per-unit effort.

Either oil contamination or increased salinity in phosphate-rich basins will allow blue-green algae to become dominant. In either case, the system could be permanently changed, even if the stressor is later removed, because if sulfate is present in quantity, a blue-green algal mat will cause very large diurnal fluctuations in the H_2S and O_2 profiles. (There are vast areas of the Persian Gulf covered by such mats.) The high H_2S concentrations of such lagoons and basins make them devoid of macroscopic animal life. Just about the only organisms able to survive under those conditions are blue-green algae (by virtue of their ability to carry out both oxygenic and anoxygenic photosynthesis), certain photosynthetic bacteria, and sulfate-reducing bacteria. The system is still highly productive, but now 99% of the production is utilized by sulfate reducers (the rest sediments out). Here we have an example of an extreme environment that is actually full of life.

Even if lagoon ecosystems become dominated by algae, these ecosystems are still used by many cultures as sources of salt. And even more importantly, such salt-evaporite basins, after being buried and compressed, give rise to salt domes, which are associated with two thirds of the oil in Texas and Louisiana. The oil itself is derived in part from the algal mats previously mentioned. These are highly significant deposits not only of oil and salt but also of elemental sulfur. (Certain biogenic deposits of sulfate minerals are also associated with fossil algal mats.) In geologic eras when shallow seas were common, this transformation may have played a key role in the long-term temperature homeostasis of the planet (via atmospheric CO_2 removal).

Other Halophytes

We have already visited the extreme deserts of the Pacific coast of South America. A very successful plant family found in those deserts is the Nolanaceae. This family is endemic to those deserts and is among the most halophytic of all plant families.

Tamarisks (*Tamarix* spp.) belong to a family of trees and shrubs (Tamaricaceae) that evolved in the salt deserts of central Asia. Like many other plants of those barren lands, some of the tamarisks have become weeds over most of western North America and Australia. They are shrubs of salty water courses or of salty springs. Their habit of transpiring large amounts of water once their roots reach the underground reserves makes them a formidable adversary of the plants that compete with them. Tamarisks can not only depress the water table beyond the reach of other plants but also concentrate the salts, making the surface waters too salty for both plants and animals.

I must admit a personal affinity for the last halophytic habitat I will talk about—the mangroves. These places, technically called mangals to distinguish the place from the mangrove tree, are at once mysterious and threatening, yet peaceful and serene. I have made a small exception and included these places in a book about limiting environments because mangroves highlight a curious limitation to woody growth. As we travel north or south past their cold-tolerance limits, the low-energy (protected) coasts revert to salt marshes. There are other trees that can grow using fairly salty water (among them, the tamarisks we have already discussed), but none can withstand the conditions in anaerobic salty mud better than mangroves.

The mangroves are a heterogeneous group of trees. Their countenance differs drastically, depending on the species, and so does their ability to withstand salinity, anaerobic muds laced with high concentrations of hydrogen sulfide, and the unpredictability of the waves. All mangroves are basically tropical plants; and in the tropics, especially in southeast Asia and Australia, they have developed into a multitude of species belonging to several families.

Atlantic shores can only boast four species of mangrove. In the tropical Atlantic, the red mangrove (*Rhizophora mangle*) is the quintessential mangrove tree, whereas in the Pacific, its relative *Rhizophora stylosa* must share the arena with several very highly adapted members of its own and other families. There are thirteen or so genera of mangroves in the western Pacific alone.

The western Pacific region is the point of origin for two of the most widespread mangrove genera, *Rhizophora* and *Avicennia.* The several species in these two genera range over both the Pacific and the Atlantic. *Avicennia* in particular has expanded far from its land of origin and can now be found along the northern coasts of New Zealand as well as the coasts along the northern Gulf of Mexico. The limits of the distribution of this most temperate of mangroves can be correlated with air and water temperatures, but whether the boundary is due to a lack of summer heat or the destructive influence of winter cold is not known. It is clear however, that in northern Florida (the North American northern limit) devastating freezes every five to ten years decimate the local populations.

6 Seasonal and Vernal Wetlands

In this chapter we will take a look at the life that lives in and around temporary fresh waters. A whole book could easily be devoted to this interesting topic, but here we will limit ourselves to just a few examples to demonstrate the intrinsic harshness of these environments.

Water at infrequent intervals can make a region alternate from desert to shallow lake or wet savannah and back again. The challenge of such places lies not in extreme environmental conditions of drought or anoxia, but in the sometimes-conflicting demands they make on living things. Depending on the length of the aquatic phase in these wetlands, some aquatic organisms can finish a complete life cycle, and some terrestrial organisms can do the same at the periphery of the inundated area. Organisms survive the dry phase of this boom-and-bust cycle either by storing water, by tapping stored water, or by producing special dormant structures. When the oscillations are predictable, organisms have more time to prepare, and they adapt through specially constructed dormant structures such as the tuns of water bears and the resistant eggs of phyllopod crustaceans. When the environment oscillates from wet to dry unpredictably, some organisms adapt through anhydrobiosis: the almost complete cessation of all living processes following extreme dehydration. In invertebrates, resting eggs or cysts are found in a variety of groups, from water bears to tadpole shrimp. Vertebrates are not as resistant, yet entombed in dried mud, some toads and lungfish can spend several seasons waiting for rain. Among multicellular organisms, vascular plants take the grand prize for the longest dormant stage. Some seeds can germinate after many hundreds, if not thousands, of years of dormancy. Some of the most outlandish claims are highly contested, but the lesson is clear—if you package it well, it will last.

Definitions

In our discussion of seasonal pools we will include pools or ponds that, though present from year to year, do not hold water long enough to allow a typical pond or marsh ecosystem to develop. (This approach fits well with this book's goal of exploring limiting environments.) These pools or ponds are shallow enough to be dry for some part of the year, yet deep enough to have typical (albeit quickly reproducing) aquatic organisms inhabiting them in the wet season. Since we dealt with saline and hypersaline waters in the last chapter, the seasonal waters we will concentrate on here are nonsaline.

The broad class of temporary pool habitat that we have defined can be broken down further into two subclasses. The first class consists of those pools that hold water for a very short time. In desert environments these pools exist for an insufficient amount of time for vascular plants to develop, but algae and bacteria do grow. In wetter climates such pools are occupied by typical upland vegetation, and they are not as biologically challenging as the next class of pools. The second class of temporary pools consists of depressions that hold water long enough to inhibit the growth of typical upland species that might germinate in the transitory moist phase. If typical upland and wetland plants cannot grow in such environments, then, what does grow? Plants and animals with several contrasting strategies grow in such environments. One strategy is to accelerate all activity to avoid the unfavorable stage of the wet-dry cycle. Another strategy is to go dormant during the unfavorable phase. Some organisms use one or the other of the previous strategies while a few avail themselves of both. Yet a third strategy is to physiologically adapt to the change and survive both phases, albeit with decreased growth in each. We will examine each of these strategies in turn and meet the organisms that use them. It should be clear by now that *pool* is a misnomer for this type of environment, since by no means is all the biological activity confined to the wet phase. Of course, organisms that grow and reproduce in a seasonal environment require water, but the pool stage of this environment might be too wet for them. These organisms must conduct their business after the pool recedes, but while there is still sufficient moisture in the soil.

African Ephemeral Rock Pools

The race goes to the quick; when water is an ephemeral commodity, good timing and speed are among the best adaptations. In Australia, a few mosquitoes can complete their aquatic phase in as little as three days. This scenario repeats itself in many parts of the world, but in at least one particular case the insects involved do not rely just on their speed. The larva of the tube-building chironomid midge *Polypedilum vanderplankii* lives in ephemeral rock pools throughout tropical Africa, from around Lake Chad in Nigeria to around Lake Chilwa in Malawi. Algae can exploit these rock pools even if they last only a few days to a few weeks, and once algae find them and detritus starts accumulating, it is only a matter of time before this midge finds the algae and lays its eggs. The pools this midge colonizes are typically 2-3 cm deep and 1-2 m^2 in area. There is only one hitch—these rock pools have a high probability of drying out before the midge larvae can complete their life cycle. The larvae's solution is to dry with the pool. In the dry state, these larvae can survive the searing temperatures on the surface of the dry sunbaked rocks. When the pools fill again, the larvae will take a few more steps on their spasmodic journey toward adulthood.

The African rock pool environments we are discussing are more complicated than the previous discussion suggests. At least in Malawi, there is another midge that inhabits pools similar to those inhabited by *P. vanderplankii*. This is the biting midge *Dasyhelea thompsoni*. This midge seems to occupy pools that are one tenth the volume of those occupied by *P. vanderplankii*, but curiously, these pools dry out on average at the same rate as those of *P. vanderplankii*. For both these midges the monthly mean durations of their watery worlds varies between 2 and 5 days. This biting midge uses a completely different strategy to survive the drying of its pools than that used by *P. vanderplankii*. Each of its larvae constructs a watertight capsule where it aestivates during the dry periods. The mystery is why these two species seldom coexist in the same pool.

A few of the pieces of this puzzle are now known. For starters, *P. vanderplankii* is a poor colonizer of newly filled pools, whereas the biting midge quickly finds and lays its eggs at the edge of such pools. The females of the biting midge seem to be drawn to pools with large fluctuations in water level. Another cue used by the ovipositing females seems to be odor, since biting midge larvae are found soon after a pool first receives the fresh scat of civet and genet cats. Once the eggs hatch in the chosen pool, a whole new set of problems have to be overcome by the larvae. The most important of these seems to be finding appropriate aestivation sites. The biting midge larvae can successfully colonize only ephemeral pools where the rock surface is rough enough for the larvae to find protected aestivation sites. Without such sites, the high temperatures of the dry rock surfaces can easily kill these larvae.

On rare occasions, these two species of midge have been found coexisting in the same pool. That it doesn't occur more often may be a sign of intense direct competition, or more likely, the presence of fresh civet scat may change the chemical composition of the water to the detriment of *P. vanderplankii* but to the benefit of *D. thompsoni*.

In pools that last slightly longer than a couple of days, the invertebrate inhabitants have enough time to make almost indestructible cysts that can come back to life once the water returns. The most famous and perhaps the best at this strategy is the little brine shrimp that we encountered in the last chapter, so well-known to children by its commercial epithet of sea monkey. Other fairy shrimp as well as other crustaceans are equally capable. Because of this ability to produce resistant cysts and eggs, some of the most ubiquitous inhabitants of short-lived temporary pools are aquatic crustaceans. Of these, the following five groups are the most common: tadpole shrimp (Notostraca), fairy shrimp (Anostraca), clam shrimp (Chonchostraca), seed shrimp (Ostracoda), and copepods (Copepoda). Other small animals include rotifers, nematodes, and a variety of insect larvae.

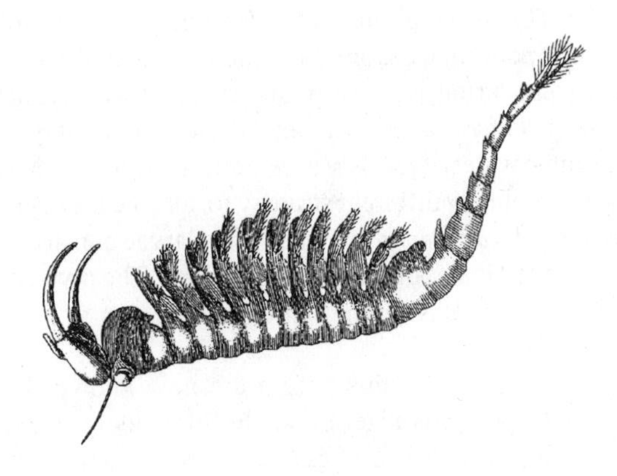

A fairy shrimp in its characteristic upside-down swimming posture.

The Stages of Longer-Lived Pools

Since the biological activity of these temporarily wet environments is not limited to the wet phase, we will follow the suggestion of P. H. Zedler and consider four different phases in the seasonal development of the longer-lived pools. He named and described those phases for vernal pools, which are a particular subset of seasonal pools that fill in the winter and dry by late spring (hence the name vernal). Vernal pools are found mainly in Mediterranean climates, in which winter is the wet season and there is a summer drought. Since much of the information available on seasonal wetlands is specifically about vernal pools, they will be a major part of our discussion. There are definite similarities between the vegetation of a vernal pool and that found on the shore of a fluctuating lake; however, the vernal pool is completely dry during the unfavorable season and water, if available, may be far underground, or under an almost inpervious layer of clay.

Wetting Phase

The wetting phase starts as the first rains moisten the ground and proceeds until the pool starts to fill. Then, the water level may become too deep for some seedlings.

Aquatic Phase

The plant species of this phase are characterized by fast generation times and tough propagules. Some wetland species manage to grow, but they must also have tough perennating structures like bulbs or rhizomes.

Drying Phase

The high moisture content of the soil allows those seedlings that survived the flooding, either by being some distance from the center of the pool or by going dormant during the worst of the flooding, to now grow and flower. The aquatic animals must now disperse or aestivate.

Drought Phase

Most activity ceases, and only perennating structures of plants, seeds, and cysts are left in the cracked earth. Some plants can still tap the now deeply buried wetting front, and they can flower for months during the worst of the dry season using this hidden source of water.

This four-phase sequence has been suggested for those seasonal environments traditionally labeled as vernal pools, but it also applies to more fleeting temporary pools.

Physical Setting

Several factors conspire to make a temporary pool. One factor is the seasonality of the rainfall pattern. Other factors include the topography and the hydraulic properties of the soils. For the pool to form, the substrate must be relatively impervious to water; a substrate such as bedrock needs no help in achieving this state. Alternatively, some pools overlay soil rich in clay. Often the clay is of an expansive type that, when wet, swells enough to seal the bottom of the pool. Therefore, once the first rains fall, the surface of the ground becomes impervious to further water penetration and the pool quickly fills. The water in these pools has only three escape routes: through evaporation, through percolation into the subsoil beneath the pool, or through overflow. Since we are dealing with freshwater pools, evaporation often is not the main avenue of water loss, otherwise the salinity of a pool would increase as rain washes salts from the surrounding countryside into it. California's vernal pools, for instance, retain the same low salinity year after year; this implies that there is mass flow of water under them, or that they overflow often enough to be flushed out.

Aquatic Phase Organisms in California's Vernal Pools

Vernal pools start to fill when their clay bottoms seal after the first rains; at this time a few plants germinate and become the dominant vegetation in them. In California, some of the most important of these plants at this stage are the quillwort (*Isoetes* spp.), the spike-rush (*Eleocharis* spp.), and the water-starworts (*Callitriche*). The first two are perennials with a rush-like growth habit, but spike-rushes are flowering plants, while quillworts are normally grouped with some of the most ancient vascular plants in the Division Microphyllophyta. *Isoetes* is thus more closely related to the club mosses that have been around since the early Devonian period (400 million years ago) than to the flowering plants. As in the aeolian zone of high mountains, the ability of ancient organisms to deal with modern extreme environments can also be observed in vernal pools. P. H. Zedler states:

> Vernal pools are another special habitat in which plants built to an ancient design can still persist, because no better solution has been found to the problems of surviving there than those first perfected in the coal swamps. (Zedler, 1987)

Isoetes is present in other habitats around the world, but it is not common, except in seasonal pools. It, like so many other wetland plants, within its leaves has tissue with hollow spaces (aerenchyma), but it also has a somewhat less common trait; it has hollow spaces in its corm (bulblike underground structure). A few other plants, including the water-hemlocks and angelicas of the carrot family, also have spaces in their corms. All species of Isoetes are also physiologically well-prepared for the seasonal pool environment. They possess a form of crassulacean acid metabolism (CAM), the same type of metabolism we encountered in many desert succulents. CAM allows desert succulents to obtain carbon dioxide at night—when it is less costly in terms of transpired water—and save it in the form of malic acid and other organic acids. During the day, to feed photosynthesis, carbon dioxide is slowly released from its internally stored form, and there is no need for the plants to open their stomata. The role of CAM in the mostly submerged *Isoetes* must be different from its role in desert succulents, since during the growing season water is not a limiting factor in vernal pools. J. E. Keeley has proposed that stiff underwater competition for CO_2 is the main force driving *Isoetes* toward CAM. He measured wide pH fluctuations in temporary pools. Since high pH is linked to the unavailability of CO_2 in the water column, those fluctuations tend to support his claim. Other supporting evidence is the lack of CAM in leaves that develop as the pools dry.

In seasonal wetlands of the high Andes, *Isoetes* plants have dispensed with stomata altogether. Somehow, they obtain all the carbon dioxide they need from the anaerobic sediment present when the pools flood. With no stomata, these *Isoetes* can survive the dry season with very little water loss. This improved water conservation helps the Andean *Isoetes* to be evergreen, whereas the species in the vernal pools of California face the summer drought by letting their leaves wither away.

For fertilization to take place, *Isoetes* plants need a surface with a liquid film, as do all other seedless vascular plants. In the Californian *Isoetes*, the sporangia of the plant are located in old leaf bases; some of these contain female megaspores while others contain male microspores. Once those leaves decay, the sporangia are exposed to the environment. This exposure triggers the development of the female megaspores, which split open, revealing several archegonia. The male microspores then produce spermatozoa that make their way to the archegonia and fertilize them. The embryo that results is the next *Isoetes* plant (the sporophyte). *Isoetes* has thus minimized the role of the gametophyte, but it still depends on the presence of a liquid film to effect fertilization.

Other aquatic-phase vegetation in vernal pools includes the ferns *Marsilea vestita* and *Pilularia americana.* Unlike most other ferns, which develop sori (clusters of sporangia) underneath their leaves, *Marsilea* develops its sori inside a larger structure, known as a sporocarp, that is found just below the soil level. The sporocarp is relatively drought tolerant, and it can survive passage through the gut of small animals.

Other plants of the aquatic phase include species of the genus *Callitriche.* The water-starworts (as they are called) are rooted, and they can be totally submerged or partially floating plants. They display aquatic heterophylly; that is to say, they have differing leaf shapes depending on whether the leaves are below or above the water line. The floating leaves of *Callitriche* are relatively broad and short and form a star-shaped pattern, while the submerged leaves are narrow and long. *Callitriche* plants have both male and female submerged flowers; at least in some species, the pollen grains germinate inside the male flower, and the pollen tube makes its way through the vascular tissue of the plant until it reaches the female flower. It has been proposed that some *Callitriche* seeds are produced through apomixis; this means they develop without fertilization. Here, as in the High Arctic, apomixis is a means of assuring that some seeds are produced even in short and unpredictable growing seasons. Another adaptation of this plant to the variable environment of vernal pools is its seed-planting behavior. After fertilization, flower pedicels are stimulated to curve downward and eventually penetrate the soil beneath the plant.

Another plant of the aquatic phase with a curious sex life is the flowering quillwort (*Lilaea scilloides*). This plant produces emergent and submerged flowers. The emergent flowers can be either perfect, or just male or female, but the submerged ones are always female. These submerged flowers stand no chance of encountering pollen where they are. Their solution is to produce a very long style, like the silk of maize. As anyone who has seen the sheen of pollen on the surface of ponds in spring can attest, the surface film can act as a trap, and styles floating in this milieu may have a better chance of cross-fertilization than those suspended in midair.

Spike-rushes sometimes dominate the larger, longer-lasting vernal pools, making them more like vernal marshes. Spike-rushes spread by rhizomes to form extensive clones, but like other sedges, they also have wind-pollinated flowers. In some pools where the water level is particularly high, spike-rushes might spend most of the growing season submerged. In such circumstances they may be starved for carbon dioxide, as is the case with *Isoetes*. This has been put forth as the explanation for the spike-rush's switch to C_4 metabolism under these conditions. Other plants, including the minuscule waterwort (*Elatine*), manage to spend their whole annual cycle, from seedling to seed, submerged. *Crassula aquatica*, a diminutive member of the stonecrop family (Crassulaceae), is another aquatic-phase annual that can deal with submergence because of its CO_2-uptake mechanism (in this case, the CAM pathways).

The harsh and unpredictable environment of seasonal pools has produced plants that rush to create propagules instead of putting too much effort into sexual reproduction. They also seem to produce multitudes of seeds, but with few means of dispersal. Producing large numbers of seeds in a variable environment makes sense, but not distributing them far and wide might seem like folly. However, animals are one of the best mechanisms to get seeds from one hospitable pool to the next, and it might be that dropping a large crop of seeds onto the drying mud is one of the best way to attach them onto passers-by. Consequently, selection pressures may favor tiny seeds delivered straight onto the surface of the mud.

After the pool starts to dry, the aquatic plants die or go dormant while a new set of plants, amphibious ones, becomes dominant. Some of these plants change from aquatic to desert plants in a remarkable transformation that brings to life the dichotomy of seasonal wetlands. One such amphibious plant that beautifully displays this transformation is the coyote thistle (*Eryngium* spp.). The young coyote thistle plants may contend with complete submersion during the aquatic stage, when they look like soft rushes, even having aerenchyma-like tissue in their leaves. As they do in true rushes, these spaces presumably lower the resistance to oxygen diffusion, and they also serve as a means to get rid of the volatile toxic products of anoxia. During the drought stage such a low-resistance diffusive path to the atmosphere becomes a handicap, but this plant has adapted by leading a double life. As the pool dries, new leaves slowly turn grayish-green, become lobed and eventually prickly, and flatten out as the water recedes. By the time the water is gone, the coyote thistle has become a well-adapted desert plant. Coyote thistles are common vernal pool inhabitants, and they are some of the few perennial dicots to inhabit these pools.

It seems that the best way to survive the harshness of the summer drought is as a seed; therefore, the most successful vernal pool plants are annuals. Amphibious annuals abound in vernal pools; they include several species of *Downingia*, a small plant with showy flowers that many times germinates before the pool fills. *Downingia* displays aquatic heterophylly. These plants require the water level to drop below their growing tip in order to bloom; they can cover the edge of the pool with white, blue, or yellow flowers from March to April. These plants produce dust-like seeds that are released onto the ground as their capsule disintegrates. One of the most common annuals in vernal pools are the small composites in the genus *Psilocarphus* known as woolly-heads or woolly marbles. These plants have an amphibious habit, as *Downingia* does, but they can withstand drier conditions and can be found growing in muddy depressions that do not hold standing water during the wet season. Like the seeds of other annuals we have discussed, the seeds of *Psilocarphus* are retained after the death of the plant, and they are covered in a woolly coat that may serve to protect them from heat. This use of woolliness is the opposite of what has been proposed for composites growing on high mountains and in the Arctic; not all coats are created for the same purpose.

Some amphibious species such as the coyote thistles may be active into the summer, but for the most part the species that make the summer season their own are not confined to vernal pools and have more widespread distributions. These species include members of the spurge family, which are common in the surrounding grasslands. One of these spurges is dove weed (*Eremocarpus setigerus*), which starts to grow in earnest once the water has receded. It grows well into late summer or early fall, when it sets seed and dies. Other native annuals known as tar weeds are also quite common in and around vernal pools. They possess some tolerance to flooding, and with their amazing ability to withstand heat and drought, they sometimes become the only green things in vernal pools by early summer.

Some cosmopolitan annual weeds have also found vernal pools to their liking. The stork's bill (*Erodium* spp.) is an example of an introduced annual weed that has become a common member of the vernal pool community. This plant has what amounts to a spring-loaded seed that works its way into the soil with every exposure to wetting and drying—one of the reasons it has become so common in temporary pools.

The very short nonvascular plants that inhabit vernal pools might be missed on first inspection, but they are quite common and may form continuous carpets of green between the taller vascular plants. Among those nonvascular plants are mosses and liverworts. Mosses of arid regions are especially well-adapted to the pool's edge and recover from prolonged periods of almost complete dehydration once the pool starts to fill. Whereas all mosses are perennials, some liverworts are annuals and survive the dry season as spores. The perennial mosses and liverworts dry up entirely and await the return of the rains.

Inhabitants of Other Temporary Waters

Larger animal inhabitants of temporary pools are confined to those pools that last more than several weeks. These animals include several types of amphibians; notable among those in North American pools are the desert spadefoot toads of the genus *Scaphiopus*. These toads aestivate for up to several months in burrows up to 1-meter deep, waiting for the rains. Because of their accessibility, these toads have been the subject of thorough investigations into the adaptations that allow them to survive in a semidesert environment.

Unlike most other vertebrates, many amphibians can absorb water through their skins. This ability is a two-edged sword, however, and in most amphibians in contact with dry soil, water leaves the animal's body along a water potential gradient. To counteract this, Couch's spadefoot toad (*Scaphiopus couchi*) has the remarkable ability to concentrate urea in internal tissues to high levels. This internal urea solution allows the toad to draw water from soil with water potentials of -10 to -15 atmospheres. The water potential of soil has several different contributing factors, depending on the type of soil and the salt load in it. If the soil is not salty, most of its water potential can be accounted for by the *matric potential*; this is the force that holds water on the surface of soil particles, and it increases as the soil moisture decreases. A spadefoot toad buried in sandy soil can gain water from it until the soil's water content drops below 3%. Once the rains come, the sound of raindrops hitting the ground causes the toad to stir to life. Once they emerge, spadefoot toads seek the nearest body of water to mate there. The females lay eggs that will hatch in one or two days, thus starting the tadpoles' race against time. The tadpoles of these desert amphibians are some of the fastest-developing amphibian larvae, being able to complete their development in under two weeks.

Semipermanent Waters

When temporary pools are generated by rivers that overflow their banks, they become semipermanent, but there is still the chance that during a drought, they will dry completely. Such semipermanent waters can be found throughout the world, but in South America, Africa, and Eastern North America these waters are occupied by remarkable creatures. The plants that thrive in such environments are typical wetland species that possess some drought-avoidance mechanisms. The animal inhabitants of these semipermanent waters include several lungfish species (order Dipnoi). Lungfishes are ancient creatures, dating back to the Silurian period (400-440 million years ago). They form one branch of the subclass Sarcopterygii. The other branch is occupied by the lobe-finned fishes, which can be further divided into the extinct rhipidistian fishes (that gave rise to terrestrial vertebrates) and the Coelacanthoidei, whose sole living member is the coelacanth. There are three genera of lungfish still living: one in each austral continent except for Antarctica. Several species of African lungfish, as well as the South American lungfish, live in stagnant semipermanent waters and can aestivate. Judging from the finds of fossilized lungfishes still in their cocoons—fossils dating from the Triassic period (180-225 million years ago)—their ability to aestivate is an ancient one. (It seems that, at least in the Triassic period, semipermanent stagnant pools were a common occurrence, and maybe they were an impetus for the evolution of terrestriality in vertebrates over 100 million years earlier.) The morphologically primitive Australian lungfish seems to have lost the ability to aestivate and is found only in several permanent rivers in Queensland. In preparation for aestivation, lungfishes burrow into the soft mud of the drying pool, where they hollow out a cavity and secrete copious amounts of mucus. The mucus eventually dries and forms a barrier to moisture loss. While the pool dries, the lungfish keeps open a breathing orifice through its cocoon and up to the surface of the mud. Lungfishes have internal gills, but depend almost exclusively on atmospheric oxygen and their simple lungs to survive in their stagnant homes as well as in their cocoons. Once it has entered aestivation, the lungfish constrains its metabolic activity to the bare minimum. It does not excrete any waste products, and those that would normally be voided in its urine build up in its bloodstream. Its oxygen consumption drops to one third of its normal rate, and the fish may lose 30% of its body water.

A similar case is that of the sirens (family Sirenidae), strange neotenous salamanders unique to southeastern North America (see figure 1). (Neoteny can be defined as normal development of reproductive potential while other adult traits are arrested.) Figure 2 shows the largest of these salamanders, the meter-long greater siren (*Siren lacertina*). There are several other neotenous families of salamanders, including the mud puppies and the amphiumas, but the sirens are enigmatic from many other points of view. Their taxonomic relationships are difficult to discern, partly because their neotenous nature masks other characteristics. These elongated, wholly aquatic salamanders are quite adept at surviving in stagnant waters. They possess lungs and retain working gills throughout their lives. They alone, among all amphibians, have totally separate atria and also separate ventricles. This adaptation prevents the mixing of oxygenated and deoxygenated blood and assures that the maximum amount of oxygen is supplied to the tissues. At least one of the sirens, the lesser siren (*Siren intermedia*), can aestivate in a fashion reminiscent of that of the lungfish. However, they are not nearly as well equipped as the lungfish to survive prolonged drying of their stagnant homes. They also produce a parchment-like cocoon that, like in all amphibians that aestivate, is derived from shed skins instead of from mucus as in the lungfish. They can manage to aestivate for a few months in the sand or mud below a dried pond.

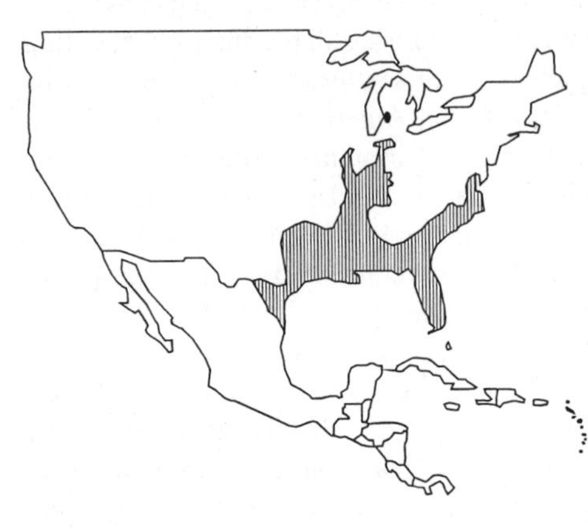

Figure 1 The distribution of the Sirenidae.

Figure 2 The greater siren.

7 Permanent Aquatic Environments

Why Are Some Aquatic Environments Limiting?

A watery world seems like a congenial environment, protected from large temperature fluctuations and desiccation. It is true that many aquatic environments are lush, and among them are some of the most productive and diverse environments on Earth. However, life in aquatic environments poses its own set of challenges: oxygen may be low or almost absent, and there might be contaminants that make the water too acid or toxic. These challenging environments are the ones we will explore in this chapter.

There is no denying that our planet is a water world and has been so since before the dawn of life. There is also much evidence that the early seas were chemically very different from the seas of today. One of the greatest differences was the absence of oxygen. The change from a reducing to an oxidizing atmosphere wreaked havoc in the living communities at the end of the Precambrian period. Selection pressures were immense, but some organisms found mechanisms to deal with the oxygen menace. Other organisms, however, simply retreated to environments that remained free of oxygen, for example, the muds beneath still waters. (These organisms currently play very important roles in the cycling of sulfur, selenium, nitrogen, and other elements on Earth.)

The transition to an oxygen-rich world may have been crucial in the evolution of multicellular life as we know it. The advantage of oxygen is that it allows organisms to carry out life at high speed. A constant body temperature (homoiothermy) is dependent on a high rate of aerobic respiration, and all mammals and birds require vast amounts of oxygen. This makes homoiothermic animals ill-adapted to environments with an unpredictable oxygen supply. Some other animals, when confronted with low oxygen levels, can switch from aerobic respiration to stopgap metabolic pathways that will take them through the difficult times. Still other animals evolved to actively seek an anaerobic retreat for part of their life cycle, since predation is slight there, and food many times is plentiful. Some of these latter animals (as diverse as worms and insects) have hit upon the concept of oxygen-carrying molecules that allow them to shuttle from the aerobic world to the anaerobic world and back again. The most popular oxygen-carrying molecules are based on iron; hemoglobin-like pigments are common in groups as diverse as sewage worms (*Tubifex* sp.) and midges (*Chironomus* sp.) that live in anoxic environments.

There isn't just one recipe for dealing with low oxygen, different animals have different adaptations for dealing with this problem. The smaller and simpler an animal is, the fewer problems it has dealing with low oxygen, but special adaptations must still be involved. In gold fish, low oxygen levels stimulate the switch to the production of alcohol as the energy-yielding metabolic pathway. This pathway is not as efficient as the production of CO_2 and water made possible by oxygen; but, since goldfish don't have lungs, they can't be too choosy. Other fish, such as the antifreeze fish that have been found under the 420-meter-thick Ross Ice Shelf in Antarctica, deal with low oxygen by carrying out life's processes in slow motion.

Modern plants depend on oxygen as much as modern animals do. Plants do make oxygen, but only in daylight and only in their green tissues. Aquatic plants, especially, sometimes run the risk of suffocation. Swamp trees such as bald cypresses (*Taxodium distichum*) and swamp tupelos (*Nyssa biflora*) create structures that help them get air to their oxygen-starved root systems. Rice plants use special conduits along the surfaces of their leaves to channel oxygen down to their roots. If the plants are totally submerged, they cannot send oxygen down to their roots; then they must resort to anaerobic respiration, just as the tired muscles of a sprinter do. Lack of oxygen at the level of the roots has other dire consequences for plants. A constant stream of oxygen diffusing out of the roots keeps toxic

Distribution of the genus *Taxodium*.

levels of iron and manganese in the soil at bay; without oxygen, plants risk being poisoned by these toxic metals.

Low oxygen levels in water can arise through a variety of circumstances. The solubility of oxygen in water varies inversely with the water's temperature and its salinity. A hot briny pool may have one third of the oxygen concentration of a cool fresh-water pool simply due to its temperature and salinity. However, the main reason for a diminished oxygen concentration in water is a high concentration of organic matter and/or a high concentration of easily oxidized inorganic compounds. For example, ferrous compounds and their biochemical oxygen demand (BOD) are a long-lasting legacy of abandoned mines. Water seeping through coal and other mineral deposits exposed by mining activities picks up reduced forms of iron and other metals. Once this contaminated water finds its way back toward the surface of the land, it trickles into surface streams. The load of reduced metals can rob mountain streams of their oxygen, especially where the contaminated water enters the stream. This depletion of oxygen by inorganic compounds is mediated by organisms to some extent, but it can also proceed spontaneously, while the depletion of oxygen by organic matter is mainly mediated by organisms. In various places on Earth these two processes conspire to rob the water column in lakes and rivers of its oxygen. The organisms that live in these waters must make periodic excursions to the surface to avail themselves of a plentiful oxygen supply, or they must change their physiology and biochemistry to deal with a whole host of anaerobic difficulties.

The Anaerobic Retreat

Once oxygen became a force in the history of the Earth, organisms found ways to use this hazardous substance to their advantage. Since most eukaryotes are dependent on oxygen for a crucial step in mitosis, oxygen may have even been instrumental in the evolution of the first true metazoans. Oxygen may thus have been the needle that allowed the weaving of such a successful thread into the tapestry of life on this planet. With very few exceptions (and none unequivocally proven), all true metazoans need to be exposed to oxygen at some point in their life cycle.

As we have seen, there are some places where the action of chemical and biological oxidations has led to the exhaustion of oxygen. These are retreats for anaerobic organisms that were very successful in aeons past but now spend their existence waiting for some celestial cataclysm that might release them from their prison, at least for a little while, until the atmospheric O_2 builds up again and drives them back to their retreats.

Our survey of anaerobic retreats will start with water-logged soils and then on to the plants and animals that live in swamps and lakes. Even though we will mention polar lakes, a detailed discussion of the Antarctic lakes of Victoria Land will be deferred until the next chapter because their extreme nature is caused mainly by their polar location. Since blue-green algae play such an important role in some of the most extreme aquatic environments, we will spend some time describing these algae and their current role in the biosphere.

Waterlogged Soils

The instant a soil is flooded, a set of biologically-mediated chemical reactions is set into motion. The specific reactions that occur in a particular soil depend on the previous history of that soil and the nature of its parent material. The soil may be exposed to aeons of rain and flooding that will leach most of the soluble elements from it regardless of its parent material. Some leached soils do not intrinsically contain many of the inorganic chemicals that become increasingly important oxidizing agents as the oxygen concentration in the soil drops. For example, when quartzitic sands become waterlogged, the only chemical reaction that occurs is the diminution of the oxygen concentration. In these sandy soils we can see the effect of a simple lack of oxygen without the complications posed by a chemically complex soil. The lack of oxygen in waterlogged sands poses some problems to the plants and animals living there, but as we will see later, the mechanisms they have evolved to deal with it are quite effective.

When chemically complex soils become waterlogged, there is a well-defined sequence of chemical reactions that proceed over time starting at the time of inundation; many of these reactions are spurred on by microorganisms that find low oxygen concentrations liberating. As the length of waterlogging increases, this set of chemical reactions produces a sequence of potentially toxic substances that plants and animals must deal with if they are to make the waterlogged soil their home. The sequence of reactions can be understood in terms of the reduction-oxidation potential of the soil (redox potential for short). The redox potential is a measure of the electrical potential energy of a solution. Like all thermodynamic measurements, it is a relative measurement. In it the zero level is the energy in a hydrogen gas/liquid water system under a predefined set of conditions (a system known as the standard hydrogen electrode). In theory, to measure the redox potential of a solution (in this case, a soil solution), one must insert a reversible electrode into the solution and complete an electrical path to the standard hydrogen electrode. The voltage that develops from this galvanic cell (battery) using the standard hydrogen electrode as the negative electrode is the redox potential of the solution. In simple terms, the solution can either donate electrons to the standard hydrogen electrode or remove electrons from it. If highly oxidizing substances are present in the solution, it will extract electrons from the standard hydrogen electrode and the voltage that will develop will be positive. The more highly oxidizing a substance is, or the higher its concentration, the higher the voltage. The reverse is also true; the more highly reducing a substance is (the more willing it is to donate electrons), the more negative the voltage will be.

A well-aerated soil will have a redox potential of about 0.6 to 0.8 volts. This value will diminish slowly as oxygen is exhausted after inundation, but when the oxygen is almost completely gone, the level will drop precipitously to another plateau. The potential level of this plateau will depend on the chemical constituents of the soil solution at that time. Although multicellular organisms are able to use only oxygen as the external electron acceptor, bacteria are not so limited.

In certain soils there may be a significant amount of nitrate present, and this chemical is almost as good an oxidizing agent as oxygen. Nitrate is used by a set of facultative anaerobic bacteria as a terminal electron acceptor in electron-transport phosphorylation (the main set of energy-generating chemical reactions within cells), replacing oxygen in this role. Respiration based on oxygen yields more usable energy than respiration based on nitrate, but the latter is almost as good, and the bacteria that use it thrive in periodically inundated soils. The sequence of reactions involved in the reduction of nitrate is more complicated than the one involved in the reduction of oxygen. And unlike the latter, the former sequence may be arrested; this occurs at the level of nitrite, if the temperature and pH are not high enough. Nitrite is a dangerous substance, and in high concentrations it can lead to the creation of nitrosamines; these substances are powerful mutagens that may inhibit the growth and reproduction of organisms in the soil. Whether increased mutation rates are found in organisms living in high-nitrite waters or soils is not known as of this writing. Nitrate is seldom a very plentiful component of the soil solution (except in places polluted by decades of agricultural overfertilization), and once it is exhausted, bacteria must turn to something else.

Many soils derived from basalt or serpentine have much larger concentrations of manganese than of nitrate, and as the nitrate is exhausted, bacteria start reducing the oxidized manganese compounds. In most soils the most common form of manganese is the dioxide. This chemical can be extremely inert or quite reactive, depending on its previous history. Whether it is an iron oxide or manganese dioxide, the more crystalline the oxide, the harder it is for bacteria to use it as an electron acceptor. Some microorganisms can even use manganese as the terminal electron acceptor in electron-transport phosphorylation. Once oxygen and nitrate are exhausted, the manganic ion Mn^{4+} is reduced to the much more soluble and phytotoxic manganous ion Mn^{2+}. Although both iron and manganese are indispensable to both plants and animals, a vast oversupply of one can block the uptake of the other. In plants, iron is necessary for the synthesis of chlorophyll, and increased Mn^{2+} in the soil causes plants to turn sickly yellow—a condition known as iron chlorosis. Another symptom of Mn^{2+} poisoning is the appearance of brown necrotic spots on the leaves. This condition cannot be prevented by increasing the soil's iron concentrations, and therefore it must be a direct effect of the increased Mn^{2+} present in plant tissues. Some flood-intolerant plants have been shown to maladaptively concentrate Mn^{2+} in their leaves while in flooded soils. Other chemical substances are also important in determining the toxicity of Mn^{2+} to plants. It seems that high concentrations of Ca^{2+} or phosphate inhibit the uptake of Mn^{2+} and can protect the plant from this ion's toxicity to some extent.

Iron is nearly always present in soils at higher concentrations than manganese, although in nonflooded soils it is present as insoluble oxides of ferric iron. After manganese is nearly completely reduced, ferrous iron starts appearing in large concentrations. The reduction of ferric iron in soils is almost exclusively mediated by microorganisms. Whereas many microorganisms use manganese in electron-transport phosphorylation and derive energy, most microorganisms that reduce iron do not seem to derive energy directly from this reduction; iron simply serves as a convenient receptacle for excess electrons. Even though energy may not be directly derived by most of the microorganisms that can catalyze this reaction, these microorganisms do benefit from the disposal of these electrons. No multicellular organism is known to use manganese or nitrate as a terminal electron acceptor, but some soil fungi may use the reduction of iron to enhance growth, at least in an indirect way.

Highly-adapted wetland plants deal with high concentrations of soluble iron in a variety of ways. Air conduits (called aerenchyma) in the roots of wetland plants allow oxygen to diffuse out of the roots and oxidize the iron before it enters the roots or, as a last line of defense, before it reaches the stele. In some wetland plants the aerenchyma may be a major line of defense against high concentrations of ferrous iron. A complementary mechanism used by plants to precipitate iron before it causes any harm involves the modification of the acidity at the root surface. The ability of oxygen to abiotically oxidize ferrous iron is highly dependent on acidity. The lower the acidity (the higher the pH), the more unstable the ferrous iron becomes and the faster it oxidizes. Some wetland plants can increase the pH on the external surface of their root system, thus increasing the rate at which ferrous iron is oxidized. What happens to these plants when the soil dries and the iron reverts to the insoluble ferric form? They become chlorotic and show signs of iron deficiency. This weakened state causes wetland plants, such as the blue flag iris (*Iris versicolor*), to succumb to competition when they grow in upland sites. Upland plants use strategies that seem to be unavailable to many wetland plants. They produce (from their roots) substances that reduce ferric iron to ferrous iron, thus making this element available for root uptake. On the other hand, most soil microorganisms secrete substances known as siderophores; these substances solubilize insoluble ferric iron without reduction and make it available for uptake. Thus soil microorganisms and plants are in direct competition for soil iron in iron-deficient soils. The production of siderophores has evolved in response to the intense competition for iron in dry upland soils.

As the redox potential drops from the iron plateau, the next ubiquitous compound to be reduced is sulfate. In soil with high quantities of iron the high concentration of ferrous iron favors the biological reduction of sulfate, since the sulfide produced is precipitated as the very insoluble ferrous sulfide. This type of soil is often found where parent material rich in iron is flooded by full-strength or diluted seawater. Many have the telltale dark gray color and hydrogen sulfide smell of this class of soils. Once the iron in the soil is exhausted, the free sulfide concentration increases to a level that is toxic to plants. Sulfide is a metabolic poison that interferes with iron-containing proteins known as cytochromes. These proteins are crucial in the set of biochemical reactions known as electron-transport phosphorylation, which produces energy out of the food an organism consumes. The mangroves growing in these soils are well adapted to deal with the high sulfide concentrations and lack of oxygen in them. When mangroves are cleared from these soils and they are drained for agriculture, they become extremely acid and unable to support plant life. The same thing can happen with gypseous soils, in which hydrogen sulfide can become a significant hazard for both plants and animals, and the reaeration of the soil can lead to such high acidity that no plants can survive.

The chemical reduction of sulfate observed in flooded soils also is a biologically-mediated phenomenon. Sulfate becomes an important oxidizing agent for a special set of bacteria. Unlike the nitrate-reducing bacteria, which are facultative anaerobes, the sulfate reducers are obligate anaerobes. When oxygen returns to the soil, these bacteria either die or are confined to the few remaining anaerobic interstices between soil particles.

When sulfate is exhausted, the only plentiful electron acceptor left in the soil is carbon dioxide. Up to this point in the anaerobic parade, the concentration of carbon dioxide has been steadily increasing, since it is an end product of all the previously discussed methods of anaerobic respiration. The bacteria that use CO_2 as the terminal electron acceptor are known as the methanogens, and they are unlike any other sort of bacteria on the planet. They are so different that they are now grouped in a Domain of their own—the Archaea (from the Greek *archaios* 'ancient'). They are remarkably different from most other prokaryotes. For starters, their cell wall is not made of peptidoglycans as in most other bacteria. (A few other bacteria also shun peptidoglycans; we met some of these, the halobacteria, in chapter 5.) If this were all, then we would say that the methanogens are just a group of highly specialized bacteria, but their peculiarities know virtually no bounds. These bacteria lack any sign of the normal cytochromes possessed by all free-living organisms. These proteins are so important and so highly conserved in evolution that the cytochromes from our cells are closely homologous to the cytochromes of yeast. A mutation in the genes coding for this set of proteins is an overwhelmingly bad event, because this set of proteins controls the transfer of electrons during electron-transport phosphorylation. Without cytochromes, this basic energy-transduction mechanism is nonexistent. Methanogens use a set of unique proteins to shuttle electrons between the hydrogen they use as a source of electrons and the carbon dioxide that finally accepts these electrons.

Methanogens are ubiquitous in the anaerobic areas of our world. They have even been found deep within the earth in permanent aquatic environments unlike any other--deep aquifers. In these places, methanogens seem to derive the hydrogen they use from geochemical reactions. For this reason these ecosystems may be totally independent of photosynthesis or its byproducts. The study of these environments may yield clues regarding the existence of life in outwardly barren worlds (see chapter 11).

Higher Plants and Animals

As in our survey of other environments, in our survey of aquatic environments we will proceed from the poles to the equator. Submerged aquatic environments are the only ecosystems to speak of in the dry regions of Antarctica; this may seem a strange statement until you realize that for any desert, life will revolve around whatever sources of water are present. The submerged aquatic environments provided by the Antarctic lakes have received a lot of attention. High Arctic lakes, on the other hand, are more plentiful, but have been much less studied. All these lakes show how low temperature and a very restricted light regime affect the survival of aquatic plants and animals. North of latitude 70°N the only submerged flowering plant is the mare's-tail (*Hippuris vulgaris*), with species of calcareous algae (*Chara* spp.), commonly known as stoneworts, taking up the role of macrophyte where vascular plants are absent.

Emergent Arctic ecosystems abound even in the Arctic semideserts. In these environments, the dominant emergent species are normally graminoids such as the grass *Dupontia* sp. and the alpine foxtail (*Alopecurus alpinus*) found in the American Arctic. The western American Arctic also has the cottongrass *Eriophorum triste*. In the eastern American Arctic, the ecological role of *Eriophorum* is played by another sedge, *Carex stans*.

Wetland plants differ widely in their ability to withstand anoxia. Although in nonflowing wetlands anoxia is the most obvious stressor driving adaptation, some successful wetland plants deal with this threat simply by avoiding it. Avoiding anoxia is a perfectly respectable tack, and many fast-growing species opt for this alternative. The only other option is to tolerate the anoxic conditions by some combination of biochemical, physiological, and morphological modifications. Since this option must involve processes that release less usable energy than oxygenic respiration, these organisms must grow more slowly than their oxygenic compatriots.

Splitting the adaptive strategies into avoidance and tolerance is, of course, for our benefit, since some strategies stubbornly fail to fall neatly into either category. Dormancy is one of these strategies, and it is adapted by many trees in the Amazonian floodplains. These trees may be flooded for three months at a depth of 6 to 8 meters and still manage to resume their growth once the floodwaters recede. In their underwater role, these trees serve as living freshwater reefs where fruit-eating piranha escape their carnivorous relatives and freshwater sponges grow on the thin branches, as if they were underwater epiphytes. If the flood does not abate after three months, some tree species start showing signs of distress, and after six months there will be significant mortality. Some species, however, can slow their metabolic rate and manage to survive two to three years of inundation before finally succumbing. This level of resistance is sufficient for even the most-prolonged seasonal floods, but it is insufficient for the artificial floods caused by impoundment of water in reservoirs (used to store potable water and/or produce hydroelectric power). Study of these artificial environments has yielded most of the available data on the survival of large trees under complete inundation. Even though dormancy is clearly an avoidance strategy, metabolic processes still occur, albeit at a slower rate, and eventually the dormant organism must face the same challenges as the organisms that remained active.

Aerenchyma and Pneumatophores

In some trees and in many herbaceous plants, an obvious change after flooding is the elongation of the internodes: the segments of the stem between the nodes. Often this is followed by the development of root aerenchyma, with hollow cavities for the circulation of air. Some plants (for example, *Caltha palustris*) create these cavities by enlarging the

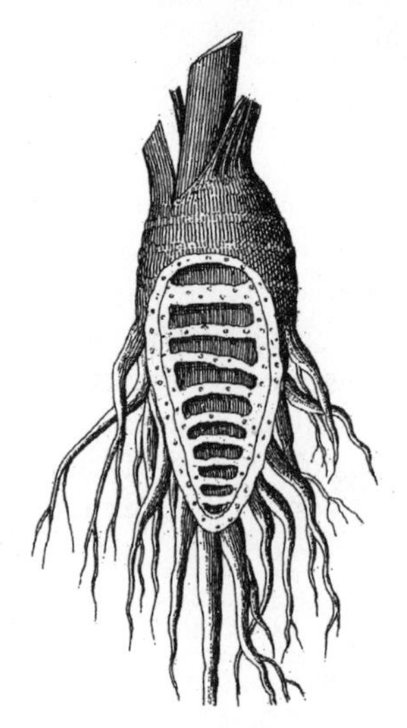

Figure 1 The aerenchyma of the cowbane.

intercellular spaces, while many others create them through cell death. Whatever their genesis, these cavities, like the tracheal system of insects, allow oxygen-rich air to reach oxygen-starved tissues. Figure 1 shows the striking aerenchyma in the cowbane (*Cicuta virosa*). The aerenchyma produced by many plants probably serves several other purposes besides oxygen delivery. Some researchers have suggested that these spaces also serve as avenues of escape for volatile toxic compounds produced by roots engaged in anaerobic fermentation. Other researchers have suggested that these spaces, which occur mainly in roots, are important for what is not there—respiring root tissues. From their point of view, the best way for a plant to maintain the same root surface for absorption and yet diminish the root's oxygen demand is to remove the central cells of each root.

The best-adapted wetland trees not only have extensive aerenchyma but also have pneumatophores: porous structures that serve mainly as respiratory organs. In wetland trees, pneumatophores are woody protuberances that rise vertically from the tree's roots. Pneumatophores provide a shorter than usual path for oxygen diffusion from the atmosphere to the root system and for the diffusion, in the other direction, of the end products of anaerobic fermentation. The pneumatophores of some mangroves (*Avicennia*, for instance) may cover the forest floor and make trudging through a dimly lit black-mangrove swamp at ebb tide one of the eeriest experiences to be had in any swamp. Another celebrated type of pneumatophore is the one found in the bald cypress and usually called a knee. Unlike the pneumatophores of black mangroves (*Avicennia germinans*), those of bald cypresses can be small knobs on the ground, or they can be the size and girth of a man. The size of the knee seems to be independent of the period of inundation suffered by the tree. Trees that spend most of their time with their roots underwater have insignificant knees, whereas those that are subjected to wildly varying water levels may have huge knees. Therefore, it could be that these knees also aid in the mechanical stabilization of these large trees; this must be advantageous in the soft muck in which some of these trees grow, especially when that muck is exposed to wildly fluctuating water levels.

Bald cypresses not only possess morphological adaptations but also have well-developed fermentation abilities. Even seasoned field biologists have been surprised by the highly active fermentation processes going on inside cypress trees. Sometimes, coring such a tree releases a burst of pressurized gas that smells just like a pile of silage. The unmistakable rancid odor is that of propionic and butyric acids. The presence of these substances suggests that the tree and the bacteria that have invaded its heartwood are engaged in anaerobic fermentation. This mixture of acid vapors is propelled by the pressure built up from the methane produced by bacteria. A bald cypress that has been flooded for a long time accumulates large amounts of organic acids and methane in its pith; these substances are produced by commensal microorganisms that normally pose no threat to the tree.

Morphological adaptations are sufficient in those cases in which the plant's roots are not deeply embedded in anaerobic mud, but more drastic action must be taken if the plant's roots delve deep into the dark mud. For instance, most species of needle rushes (*Juncus* spp.) have shallow root systems and extensive aerenchyma. Even though they are the dominant vegetation in many brackish marshes, when these plants are placed in an anaerobic chamber where their aerenchyma is of no use, they die within 24 hours. Other vegetation, similar in morphology to the needle rushes and growing next to them in the same brackish marshes, when subjected to the same treatment can easily survive for a month before taking its next breath of fresh air. Plants that show this amazing metabolic and biochemical adaptability include some sedges, cattails, and the truly aquatic members of the iris family like the blue flag.

Biochemical Necessities

On this planet, all multicellular organisms come from aerobic stock; if they have an anaerobic arsenal, it was not designed *de novo* for an anaerobic existence but resulted from modifying a preexisting aerobic repertoire. As we will see, oxygen does not simply provide a way to produce more energy than that produced by fermentation or anaerobic respiration, but it also is central to the maintenance of a dynamic internal balance: an internal *steady state*. This internal steady state evolved as the atmosphere's oxygen concentration rose. This internal balance that is so intricately linked to oxygen availability is one of the threads that tie all multicellular organisms on this planet. Multicellular organisms that have ventured into anoxic environments have had to try to maintain this balance without availing themselves of the driving force behind its creation: oxygen. Part of this balance consists of the maintenance of a pool of a high-energy chemical compound, a compound that can drive all the different processes that must be controlled or forced to go in a thermodynamically unfavorable direction. In other words, there must be some means of directing the energy available from food to the chemical reactions that keep an organism from disintegrating into a puddle of slightly salty water. All known free-living organisms have hit upon using highly labile (high energy) nucleoside-phosphate esters as the energy-carrying compound. (I've used the attribute *free-living* because the few exceptions to this rule are intracellular parasites that have lost even this basic capacity and must rely on their host's energy charge to live and procreate.) The most ubiquitous of this family of compounds are the following adenosine phosphate esters: adenosine monophosphate (AMP), adenosine diphosphate (ADP), and adenosine triphosphate (ATP). The cellular machinery is mostly powered by the hydrolysis of phosphate bonds, which are handily carried by ATP molecules. Just as the top card in a house of cards is the most unstable, the last phosphate molecule in ATP is the most unstable and releases the most energy when parted from the rest of the molecule; that is, when it is hydrolyzed. The next bond releases less energy when broken, but it is also used as an energy carrier by cells. A convenient way to summarize the energy state of a cell or an organism is by computing the *energy charge*: the concentration ratio given by (ATP + 0.5 ADP) / (ATP + ADP + AMP). The energy charge is a measure of the amount of energy left in this source. If the energy charge drops too low the cellular processes come to a screeching halt and death promptly ensues. To avoid this fate organisms must quickly adapt to the low energy charge that is an immediate consequence of low oxygen. During the first stage of anoxia the energy charge of the most sensitive organisms drops to unrecoverable levels; other more resistant organisms undergo a less pronounced drop.

In the blissful world of oxygenic respiration the cell is awash in ATP. In this world, each molecule of a simple sugar such as dextrose yields 38 molecules of ATP. Without oxygen most of this bonanza is unavailable, and depending on how well adapted to anaerobic conditions the organism is, it must make due with 1 to 8 ATP molecules per molecule of glucose. To maintain the same energy charge it had with oxygen present, an organism must either slow down all its ATP-consuming processes, or consume from 6 to 36 times its previous intake of food. Usually, increasing food intake will only lead to other derangements of metabolism such as a drop in pH; therefore, the only prudent course is to slow everything down.

For example, in eastern Asia wild goldfish inhabit stagnant waters that at times become very deficient in oxygen. As the concentration of oxygen drops, the goldfish's reaction is to increase its ventilation rate as a means of delivering more oxygen into its bloodstream. Once the point is reached where ventilation takes up more oxygen than it delivers, the fish slows its energy-consuming processes and switches to subsidiary ATP-producing chemical reactions. The fish's energy charge, although it falls as the fish becomes hypoxic, eventually recovers and stays relatively high during the rest of the anoxic period..

Another part of this internal balance is the maintenance in the cells of steady ratios of reduced hydrogen-donating molecules to oxidized ones. This is a critical part of the maintenance of the internal redox potential of the cell. Just as ATP is highly conserved throughout all life on Earth, so are the hydrogen-carrying compounds. In energy-generating pathways, the compounds in question are mainly NADH and NAD^+, whereas in biosynthetic pathways the compounds are NADPH and $NADP^+$. There is also a set of about ten indispensable coupling chemicals needed to link the different reactions inside cells to form a living, working whole. Some anaerobic multicellular organisms have been shown to produce at least eight of these. Many multicellular organisms have been successful in exploiting the anaerobic realm on a temporary basis, but none has broken the bond with oxygen completely, and I know of no multicellular organism that can spend generations in a completely anaerobic realm.

A third aspect of this complex steady state is acidity (most often expressed as its logarithmic rubric pH). Although it is biological lore that internal acidity must be maintained within narrow limits for organisms to survive, organisms are better at surviving large fluctuations in internal acidity than at surviving some other ionic changes. Aerobic respiration influences internal acidity less than many forms of fermentation do. Its acid end product—carbon dioxide—is easily removed from the internal milieu and, in any event, is a much weaker acid than most other fermentation end products. Thus, aerobic respiration produces about the most innocuous end products imaginable: water and carbon dioxide.

It has been shown that a little Indian cyprinid fish, *Rasbora* sp., can survive 100 days when kept in a sealed jar of water; such demonstrations lack scientific rigor, but they show that certain higher organisms have become masters of temporarily anoxic waters. An equally astounding feat is achieved by the common bulrush (*Scirpus lacustris*) and the saltwater bulrush (*Scirpus maritimus*); these two members of the needle rush family can maintain their energy charge for over three months, even managing to produce shoot elongation in the complete absence of oxygen. In the wild, this ability comes in handy in early spring when the rhizomes are again striving to produce shoots through the anaerobic mud. At this time, the danger for these well-adapted wetland plants is not lack of oxygen—it is starvation. During this time the plants significantly deplete their carbohydrate reserves, and if they exhaust these reserves, they will die while still buried in the mud.

We are all familiar with the buildup of lactic acid in our own muscles during bouts of strenuous exercise. In humans and in other mammals, such exertion creates an oxygen debt that must be repaid. However, lactic acid fermentation is not a very good solution to prolonged anaerobic episodes. It is a relatively strong organic acid, and the reactions that lead to its manufacture produce excess NADH and thus throw the redox potential of the cells out of balance. The goldfish's solution is to first oxidize lactic acid to the weaker acetic acid. This process yields one molecule of ATP per molecule of lactate oxidized; this yield helps to maintain the energy charge. This approach also has the advantage of allowing the restoration of the redox balance through the enzymatic reduction of some acetate to ethanol. The enzyme that accomplishes this is ethanol dehydrogenase; it oxidizes some of the surplus NADH in order to reduce acetate to ethanol, which is then excreted by the fish.

For plants not well adapted to anaerobic conditions, these conditions can lead to death through a variety of causes. When a plant switches to anaerobic respiration, it starts manufacturing toxic end products unlike the

The sweet flag (*Acorus calamus*).

relatively nontoxic end products of aerobic respiration. These end products vary with the plant and with the length of inundation, but many not-well-adapted plants produce lactic acid, whereas the better-adapted plants produce ethanol. However, plants have a harder time than animals escaping their own wastes. Although ethanol is not very toxic to plants growing under atmospheric levels of oxygen, it is toxic to plants under anaerobic conditions, and can be extremely toxic to plants coming out of an anaerobic bout.

R.M.M. Crawford, who has extensively studied the adaptations of plants to anaerobic conditions, has used pea seedlings as subjects in his studies of ethanol toxicity in plants. The only serious effect of ethanol on fully aerobic pea seedlings is a decrease in the branching of the developing root system. For plants that are grown in the complete absence of oxygen, however, the buildup of ethanol (which then cannot be oxidized and detoxified) causes seedling mortality. When a pea seedling starts producing ethanol under anaerobic conditions, it starts secreting the ethanol into the soil. If there are no air or water currents that can remove the ethanol from the vicinity of the plant as fast as it is produced, the plant succumbs to ethanol toxicity. The common pea is certainly not a well-adapted wetland plant, and although it is a convenient test subject to explore the damaging effects of ethanol, it tells us little about the adaptations of true wetland plants. In some wetland plants, for example the sweet flag (*Acorus calamus*), the rhizome is porous and lies on the surface of the substrate. This type of rhizome allows dilution and washing-out of the ethanol. The best-adapted wetland plants such as mangroves do not produce ethanol in large amounts and instead opt for organic acids like malic acid.

The Recovery Phase

Different wetland organisms have different levels of susceptibility to the toxic products of anaerobic metabolism. The more resistant organisms are therefore the more successful at growing in stagnant waters. Resistance to the products of anaerobic metabolism is not sufficient, however. When the environment improves and the organism comes out of anaerobiosis, there must be mechanisms available for it to deal with the large concentrations of toxic products that may have built up in its tissues during the anaerobic stint. However, what were waste products during the anaerobic period become good fuels once oxygen returns, and the organism can gain an energy advantage by oxidizing whatever anaerobic end products are still left in its tissues.

On the other hand, the organism may be overwhelmed when it reencounters oxygen, since its oxygen-protective enzymes may have degraded in the long days of anaerobiosis. Even if during anaerobiosis the organism maintained its levels of superoxide dismutase (SOD), catalase, and peroxidase, at the resumption of aerobic respiration the large amounts of anaerobic waste products still present in its body increase the concentration of the chemical agents known as free radicals. These agents can be produced when a reduced substance such as ethanol overwhelms the supply of oxygen. (This happens in an internal combustion engine running on a rich mixture, and it also happens in a cell running on a similarly rich mixture.)

At least some of the well-adapted wetland plants augur the eventual return to aerobiosis by steadily increasing the level of SOD. The role of this enzyme is to remove the threat from the superoxide radical: one of the most reactive of all biologically generated oxidizing agents. The way SOD deals with the superoxide radical is by making two of these radicals react together so that the electrons are no longer unpaired and the relativity decreases, at least in theory. Actually, the product of this condensation—hydrogen peroxide (H_2O_2)—can be as reactive as superoxide and, under some circumstances, can generate products that are much more reactive. In the presence of ferrous iron, H_2O_2 can split, forming hydroxyl free radicals, which are possibly the most reactive oxygen products to be found in biologically relevant settings. Hydroxyl free radicals are extremely reactive forms of oxygen that, like the proverbial bull in the china shop, can wreak havoc as they indiscriminately oxidize the most sensitive cell components.

To deal with the dangerous chemicals described above, some wetland plants have developed not only a set of tools but also the ability to prepare in advance for the coming ordeal. The yellow flag (*Iris pseudacorus*), a well-adapted European wetland species, can withstand two months of anaerobiosis without any damage. This plant prepares for those chemical thugs by steadily increasing the level of SOD (and presumably the level of the other protective enzymes as well) during the anaerobic period, although it does not need them then—truly preventive medicine! Nonwetland plants such as bearded irises (*Iris germanica*) survive prolonged periods of anaerobiosis only to succumb to uncontrolled oxidative processes once oxygen returns.

A different set of threats exists in the anaerobic soil surrounding the plants. This soil is characterized by a high moisture content, low levels of bactericidal oxygen, and high levels of microbial food (organic acids and ethanol). It is no wonder that the soil around wetland plants can be a breeding ground for organisms that, if allowed to enter a plant's cuticle, might wreak havoc inside the plant. Many nonwetland plants, including some plants that can stand short inundation periods, are initially damaged by the direct physiological effects of inundation, but what actually kills them are pathogenic microorganisms that manage to penetrate the plant's defenses once its resistance has been diminished.

Seeds as Anaerobic Packages

The coats of many seeds are only slightly permeable to oxygen and water, yet most seeds cannot germinate in a completely anaerobic environment; for example, when they are buried in anaerobic muds. Only a few grasses are known to germinate without oxygen; these include rice (*Oryza sativa*) and barnyard grass (*Echinochloa crusgalli*), a weed of rice paddies. Of the few species that have been tested, there are a few others that can germinate, albeit slowly, at an oxygen partial pressure as low as two thousandths of that found in the atmosphere. This ability seems to depend, at least to some extent, on the chemical form of the seed's stored energy. Glycogen in animals and other carbohydrate-storage sources in plants are directly funneled into the glycolytic pathways. The carbohydrate-storage sources in plants will allow a seed to survive some periods of anaerobiosis. If a seed's energy is stored in lipids, the seed will be unable to use this energy source without oxygen, and germination will have to be postponed, even if all other conditions are favorable.

Germination is only the first step toward survival in anaerobic muds. The seedling that emerges must strive toward the light, but while it is still forging through the mud, it must depend on its ability to carry out fermentation. Once it emerges into the realm of light and oxygen, it must deal with oxygen's dangerous but essential influence. This influence will shape its life above the mud's surface.

Blue-green Algae

The blue-green algae were, and still are, the masters of the lit oxygen-poor places. They, like bacteria, are prokaryotes, and many algologists prefer to call them the *cyanobacteria* to make the distinction perfectly clear. They, like few other organisms on Earth, can live in either the aerobic or the anaerobic world; they can dominate their environment precisely because of this duality. Like higher algae and plants, they have two separate systems for the collection of light energy. These two systems have been succinctly called photosystem I and photosystem II. Photosystem II is one of the features that separate these bacteria from the rest of the prokaryotic world; this invention fundamentally changed the surface geology of the planet and the course of evolution on Earth. This newfangled photosystem allowed more energy to be extracted from each photon of light, but it also had the advantage of using water as the initial raw material, instead of using molecular hydrogen or hydrogen sulphide, as is the case with photosystem I. The one drawback of using water is the ultimate production of a dangerous waste product—oxygen. Photosystem II is a masterful case of having a tiger by the tail. The stable form of oxygen is a dimeric molecule made up of two oxygen atoms in an unreactive configuration (at least at environmental temperatures and pressures). Since a water molecule has only one oxygen atom, any process that extracts oxygen from water must at some point juggle two potentially highly reactive atoms of oxygen, forcing them into binding with each other instead of with cellular constituents.

In an aerobic setting, blue-green algae have fully functioning aerobic metabolism, with all its complicated controls and fire walls that prevent the uncontrolled oxidation of other cellular structures. If the oxygen is exhausted, and there is not enough light to manufacture more, some of these blue-greens can revert to being relatively normal photosynthetic bacteria; they turn off photosystem II and go about their lives as if oxygen had never existed. A small pool in a salt marsh can be dominated by a blue-green algal mat precisely because these algae are among the few organisms that can withstand the daily oscillation between high oxygen and no oxygen.

Sediments under these blue-green algal mats are teeming with the members of the anaerobic world, while the surface of the mats belongs to the aerobic world. A blue-green algal mat is a threshold in time as well as in space; passing through it is like traveling hundreds of millions of years in Earth's history. The organisms that live in the underworld beneath the mat are prokaryotes that are either photosynthetic or nonphotosynthetic bacteria. There are myriads of the latter, but the former are specialized organisms that live on the thin boundary between the lit oxidizing world, and the dark decaying realm of the decomposers.

Blue-green algae are important even in the human sphere. They can be very useful to us as a direct source of food or as an indirect one. Other uses for them include the following: fodder for animals, a source of nitrogenous fertilizer, and a possible future source of fuel.

Blue-green algae have been used as food by several cultures at different times throughout recorded history. The Chinese consider *Nostoc commune* a delicacy, and the Japanese have used *Aphanothece sacrum*, *Nostoc verrucosum*, *N. commune* and *Brachytrichia quoyi* as side dishes since ancient times. Furthermore, in certain parts of the world blue-green algae have been, or are, a staple item of the diet, not mere delicacies. Bernal Díaz del Castillo, who accompanied Hernán Cortéz to Mexico, described in 1521 how the Aztecs living in the area where Mexico City is now "sell some small cakes made from a sort of ooze which they get out of the great lake, which curdles, and from this they make bread having a flavor something like cheese." This was probably a blue-green alga, perhaps *Spirulina platensis*, which is now beginning to be farmed again in Mexico and around the world as a cheap source of protein. At present, a related form is collected and eaten near Lake Chad, in northcentral Africa, where the alga is harvested from a corner of the lake and dried. For at least 2000 years it has been sold in local marketplaces to natives who grind it into a dust that they stew into broth and sprinkle on grain. They do eat some meat and eggs, but this blue-green-algal meal is their main protein source.

Experiments have shown that *Spirulina platensis* can be made to grow in tropical areas of the world without intensive cultivation. Under optimum conditions it will increase in weight and bulk twelvefold every 24 hours, compared to sevenfold for chlorella, a green alga. Even in colder climates, where artificial heating would increase production costs, such a protein source would probably be cheaper than anything else around.

Another aspect of blue-green algae that makes them economically important is the ability of many of them to fix nitrogen. There is strong evidence that nitrogen-fixing blue-green algae make a major contribution to the fertility of rice paddies. Since in many eastern countries peasant farmers do not fertilize their fields in any way, it appears that blue-green algae may often allow a moderate rice harvest to be gathered, when in their absence there would be a poor one. It does not seem unreasonable to suppose that millions of people survive largely because of nitrogen fixation by blue-green algae.

In underdeveloped countries, an agent that fixes nitrogen in situ has many advantages over a nitrogenous fertilizer that must be transported. The discovery of an alga-fern symbiosis was the first step toward the widespread use of such a fertilizing agent. The mosquito ferns (*Azolla* spp.) are small ferns that grow floating on water; a blue-green alga, *Anabena azolla*, lives in cavities in the fern leaves. The alga fixes nitrogen from the air and excretes nitrogenous compounds into the leaf cavities. Farmers in Thai Binh province, the most intensively cultivated area of northern Vietnam, deliberately cultivate *Azolla pinnata* in rice paddies. It has been reported that in these fields rice yields are 50 % or more higher than normal. It has long been known that *Anabena* and other blue-green algae found living free in the waters of rice paddies fix nitrogen, but the *Anabena-Azolla*-rice association was not uncovered until the seventies. Apparently, the fern provides just the right environment for algal nitrogen fixation, and the *Anabena-Azolla* combination is much more efficient at fixing nitrogen than the free-living algae.

In the next few chapters we will encounter environments where cold becomes the overwhelming impediment to life. However, not all cold is the same; in some environments such as polar lands, constantly low temperatures limit plant and animal growth, whereas in others, the limiting factors are extremely low temperatures at certain times.

8 Polar Lands

Every organism's cycles are set to its own tempo. The biological metronome does not tick in immutable increments such as hours or days. The tempo of warm-blooded vertebrates is set mainly by the size of the animal: the larger it is, the slower its tempo. For plants, the timekeepers are water and temperature; warmth allows a plant to develop at a certain rate, if it has enough water. Most plants must depend on the sun for warmth, as well as for light. When the temperature drops, plants and cold-blooded (poikilothermic) animals slow down. From the temperate forests poleward, the lack of warmth becomes an evermore severe limitation to all life; this is the subject of this chapter. The terms Low Arctic and High Arctic, which we will shortly encounter, are terms of convenience and are somewhat ill-defined. The region that extends northward of 72 or 73°N latitude to the northernmost land can be thought of as the High Arctic; the rest of the Arctic then becomes the Low Arctic.

In the polar regions, the most important limiting factor for life is a simple lack of warmth. Be it plant or animal, if an organism does not produce its own heat, it must turn to the sun for warmth. All of the arctic plants are designed to complete their development as fast as possible during the short period of constant warming daylight. Those that do not proceed rapidly on the path from flower to seed must be able to suffer no damage when captured by the stilling hand of early frosts. Since arctic plants are not capable of generating high levels of metabolic heat (we will meet some that do in chapter 9), they opt for greenhouse-like structures that allow leaves or seeds to develop faster. Nothing is without risk, however, and concentrating heat can also mean drying out faster. For plants, the interplay between the requirements for warmth and for water is one of the recurring themes of life in extremely cold environments.

Only a few warm-blooded land animals brave the arctic lands in winter. Some, like the arctic hare and the muskox, are so well insulated that they need not seek shelter during winter gales and can maintain themselves as long as there is enough vegetation exposed on wind-swept ridges. Antarctic lands are much more desolate. There are no vertebrate herbivores, and the only winter residents that do not seek the shelter of the relatively warm ocean are the southern black-backed gull, the sheathbill, and the emperor penguin.

Summer Warmth

Plants, given adequate warning, have no difficulty handling extreme cold (as we will see in this chapter and in chapter 9). Most trees in the taiga of interior Alaska and the Yukon stop growing when the day length drops to less than 20 hours, which in this area happens in late July, plenty of time to prepare for the first frosts. Why is it then that as we travel north from the taiga into the tundra these same plants disappear from the landscape? In fact, they also disappear as we climb the Mackenzie Mountains to the east. Traveling north or climbing upwards produces a decrease in the growing season. When agronomists define the growing season, they normally speak of the days between the average last frost of spring and the average first frost of fall. This applies in milder climates, but it is quite inadequate for the extremes we are dealing with here. We must distill the essence of that period we call the growing season. The key is warmth.

Heat is necessary for enzymatic reactions to proceed at a reasonable rate; and for plants as well as for poikilothermic animals, the higher the temperature, the faster development will take place. Insects that may take one season to develop at 35°N latitude may take four or more years to develop at 75°N latitude. This goes for plants also. The mastodon flower (*Senecio congestus*) is a large circumpolar daisy that often grows to one meter in the Low Arctic. In these balmy latitudes, it behaves as an annual, but it is actually hepaxanthic—it will not give up the ghost until it has successfully flowered. In the High Arctic, this plant not only is much shorter and woollier but also behaves as a biennial, since most years the first flowers are killed, and the plant must complete its task the following summer. Generation time, therefore, is not a constant for the same species in different localities.

In order to compare the ability of different species to adapt and develop in very cold environments, we need a measure that will be a constant for the same species no matter where it may grow. Since growth is a function of both temperature and time, integrating the temperature over a plant's growing period, from the break of dormancy to the setting of seed, should yield a measure that is closely related to the plant's minimum heat requirements. This measure can be called the minimum heat sum, and it is the lowest amount of heat energy a plant requires to complete its development. In practice, the heat sum is computed by summing the average daily temperatures over the growing season; the result is given in units of degree-days. When calculating the heat sum, assumptions are made as to what constitutes the growing season based on knowledge of the particular plant's physiology. Most of the time the assumption is made that plants are unable to grow when the average temperature drops to or below

0°C. This is not necessarily true, and some measurements have been made using other baseline temperatures.

An arbitrary baseline temperature might at first seem to be a critical flaw of the heat sum. An example will show this is not so. Let's take a hypothetical plant, the fastest-maturing plant in the world, and let's call it the whiz bang plant. The whiz bang plant reaches its minimum heat sum limit on the crags of the Icefield Ranges, where the temperature only reaches, on average, 1°C above zero for 30 days in midsummer. Then, at this spot, the heat sum is 30 degree-days. But on a crag just 200 meters higher lives another even hardier plant, and at this higher elevation the average temperature is only -0.1°C over the same 30 days; this plant develops in -3 degree-days. This shows that, even using an arbitrary baseline temperature of 0°C, the relative measurements reflect the developmental quickness of the plants, as long as the daily temperature fluctuations of the two environments are similar. Lower numbers represent quicker plants. Whether any vascular plant can complete its life cycle with a heat sum below zero is not known, but several can grow when their leaf temperature is -2°C.

It is sometimes easier (especially when dealing with trees) to calculate the heat sum for leaf expansion instead of the heat sum for seed production. For leaf expansion, evergreen conifers have heat sum requirements of around 950 degree-days, whereas deciduous conifers have heat sum requirements of around 600 degree-days. Woody shrubs such as willows need just over 200 degree-days for leaf expansion. Making wood slows plants down, and the farther north or the higher we travel, the fewer woody plants of any kind we will find.

There are only a few places on Earth so cold that every month's average temperature is below freezing. Generally speaking, the antarctic ice cap, the Greenland ice cap, the ice fields of southeastern Alaska and neighboring Yukon, and some of the arctic archipelagos that experience pack ice the year round, are such places. Plants do grow in the unglaciated areas that jut above the ice, but there is a terrible paucity of information on the flora-climate relationship of many of these remote places.

Warmth and Wood

In the Canadian Arctic, the 650 degree-day line translates roughly into a line where the average July temperature is 10°C. This isotherm coincides (in most places) with the transition zone between the taiga and the tundra. Some trees can grow north of this boundary. They, however, are unique, since no sexual reproduction is possible for these species north of the current tree line. Both aspen (*Populus tremuloides*) and black spruce (*Picea mariana*) have managed to grow several hundred kilometers north of the usual tree line west of Hudson Bay. These isolated groves are composed of clones of a few parent trees, relics of the last hypsithermal (the warmest period after the last ice age). These species can grow quite well at these latitudes and can even increase in number through layering. The process of layering starts when a branch of one of these trees touches the ground, maybe through the weight of winter snow or because the tree has leaned in the soggy and unstable ground. The branch takes root and sends a new vertical shoot that will, in time, become a new tree. Black spruce, since it is a tree of swamp and muskeg, forms groves of drunken trees that in their swaying attitude assure the survival of that patch of misplaced forest. These feats of asexual reproduction emphasize that, even though the process of creating wood is slow and expensive (in the currency of degree-days), the process of creating large mature seeds is even more so. Were a fire to ravage one of these clone forests, no new seedlings would take the place of these trees, for these clones have not produced mature seed in hundreds, if not thousands, of years.

The problem with seeds is that they are the end product of a long and arduous process that starts with the production of male and female reproductive structures. Then, fertilization takes place, and only then can seeds start to develop and grow. For conifers, especially, the process is quite drawn out and the sequence from pollination to mature seed may take several years. In the case of the black spruce, the trees start on their way toward reproductive structures, but no summer is warm enough to ever complete the mission, and the trees must start anew every several years without ever reaching their goal. This lack of sexual reproduction is the first limit to the growth of trees. During the hypsithermal (and also during other warm periods), clones of these trees grew even farther north; all that remains of those ancient groves are pieces of wood sunk in the soggy earth and pollen grains. The reason for their demise as the weather worsened after the hypsithermal is the second and final limit to the growth of trees: it gets too cold to make wood.

Some trees have adopted growth habits that betray their ancestry, but survival is a stronger incentive than tradition. These trees are the prostrate willows. No family of trees has been as successful in dealing with absolute bark-splitting cold as the willows. There are over 400 species of willows on the planet; together with the birches, they compose almost all of the shrubs found in the tundra. The feltleaf willow (*Salix alaxensis*) is possibly the last upright tree (or shrub) found as we travel north in the Canadian Arctic. Its last refuges are Banks Island and the southwestern coast of Victoria Island.

Beyond this point a few hardy shrubs are found up to the Queen Elizabeth Islands. However, the northwestern Queen Elizabeth Islands are too cold and dry for even these dwarf and scraggly shrubs. The arctic willow (*Salix arctica*), the most persistent of the willows, manages to survive on these islands (though only in protected sites). This willow, the arctic dryad (*Dryas integrifolia*), and the arctic white heather (*Cassiope tetragona*) form the last vestiges of truly woody growth in the Arctic. The arctic willow forgoes stately majesty for simple survival. Its stems and even its main trunk are a few centimeters in diameter and never reach more than five to ten centimeters above the ground surface. Even though wood is difficult (maybe impossible) to produce below 200 degree-days, the arctic willow stacks the deck in its favor by growing close to the ground surface where the energy harvested from the sun's rays is not lost to the cold winds. All of the tundra plants, woody or not, use this strategy. This Lilliputian approach serves the collective good of all the tundra plants by producing a frictional surface to slow the wind. Any plant that aspires to dominate its neighbors by growing above them and stealing their light finds itself humbled by the wind. A layer of reasonably still air above tundra plants produces an insulating blanket that allows leaf temperatures to climb eight to ten degrees above the temperature registered a couple of meters above the ground surface. Standard weather stations normally record the temperature of a site at a few meters above the surface of the ground, making it difficult to determine the microclimate the plants are actually experiencing.

To Seed or not to Seed

The creativity of plants does not stop at turning into dwarfs in order to avoid the chilling and desiccating winds; it also includes the creation of some wonderful and elaborate structures to help the seeds mature faster. The simplest embellishment is the fur coat approach. Some plants, such as the woolly lousewort (*Pedicularis lanata*), take this to an extreme, and in the spring look like misplaced cotton balls stuck in the ground by some mischievous child. For this approach to work, the seeds and surrounding structures must be a dark color, preferably black (so as to better absorb the sun's rays), and the woolliness cannot be so thick as to block a lot of the sun's energy. A less obvious difficulty of being woolly is a consequence of the low moisture-holding capacity of cold air. When the air trapped by the wool is heated, its relative humidity decreases greatly, or if you prefer, its moisture-holding capacity increases greatly. In order to fill this capacity, water leaves the plant part protected by the wool. Maybe because of this, some plants only insulate their reproductive structures. This is the case with the cottongrasses (*Eriophorum* spp.), which clothe their seedheads in a ball of wool where the temperature can reach 20°C above the ambient temperature. Large areas of the Alaskan north slope are covered by cottongrasses that, when in bloom, carpet the ground with cottony balls on half-meter stems. Other plants, such as the arctic poppy (*Papaver radicatum*), opt for the sun-worshiping approach, and their parabolic-dish flowers follow the sun around the horizon for the whole arctic summer. Still others protect their seeds inside structures made from translucent bracts or fused petals; among them are the bladder campions (*Melandrium* spp.), whose flowers look like miniature Chinese lanterns. The size of these greenhouse-like structures is taken to an extreme, not in the Arctic, where there is hardly enough room in the still air layer close to the ground, but in the Himalayas. There, the Brahma Kamal (*Saussurea obvallata*) uses three or four modified leaves to create an almost transparent globe encasing the dark-purple flower heads in its embrace. The whole flower structure looks like a hollow cabbage stuck on a 70-cm stem.

A few arctic and alpine plants use a different strategy. Instead of creating elaborate structures in order to speed the development of their seeds, they can arrest the development of their flowers and seeds based on the onset of frost, and they resume that development the following year, without damage to their reproductive structures. These plants have been called aperiodic because they do not use day length to time the cessation of growth. Of course, the distribution of such plants cannot be predicted based on the heat sum at a locality. Other factors, such as winter drought, might be responsible for the limits to their range. Such metabolic plasticity has been shown to occur in the mustard, dock, pink, and rose families, as well as in graminoids (grasses, sedges and rushes).

Seeds are wonderful packages; they are travelers in time as well as in space. They hold the same genetic information as, but have few of the vulnerabilities of, metabolizing tissue. To conquer new territory fast, a plant's seeds must be efficiently borne by wind and water or, if less efficiently carried, be longer-lived. The fluffy seeds of willows can travel great distances in the wind, but they live only a few weeks. Even though the level of metabolism of seeds is very low, they still have to maintain their pianissimo symphony of biochemical reactions. The heavier a seed is, the more stored energy it contains, and the longer it can stave off entropy and death. Many seeds have enough reserves to survive decades in the soil; some can survive much longer. Heavier seeds take more degree-days to mature; therefore, such seeds are mostly found in the Low Arctic. The second secret of seed success is packaging. If the seed is encased in an impervious coat, it will be able to control not only the penetration of oxygen and water, but also that of dangerous microorganisms. The most astounding report of seed longevity concerns the arctic lupine (*Lupinus arcticus*). In the Yukon, a few seeds of the arctic lupine were found three to six meters below ground, in fossilized burrows of collared lemmings. Out of the few seeds that were collected, six germinated, and one of those flowered. The age claimed for these seeds was 10,000 years. Even though this claim has been brought into question (because the seeds were not dated directly), it is likely that arctic-lupine seeds can remain viable for long periods after interment in permafrost. There are probably other frozen seed caches of other animals that will, in time, be exposed. Might not one of them contain viable seed of a species not seen on Earth for tens of millennia?

Sometimes plants must give up on seeds altogether. Both plant and animal species give up sexual reproduction at the limit of their endurance at the limits of their geographical range. In the case of many plants, clonal (vegetative) reproduction is already a regular means of gaining ground against competitors. Some imported weeds have traveled across the North American continent mainly by stolons that root and break off. An example is the water hyacinth (*Eichhornia crassipes*), which I see blooming outside my window as I write. Certainly plants in the Arctic can use these same mechanisms, but some have gone much further. Instead of relegating asexual reproduction to a secondary mechanism, they have made it their primary means of reproduction. However, strict reliance on asexual reproduction jeopardizes the species' ability to adapt to unpredictable circumstances. The spiderplant (*Saxifraga flagellaris*) reproduces by creating stolons with tiny replicas of the plant attached at the end; the whole thing resembles a miniature version of a strawberry plant. Once the stolon lands in a favorable spot, the daughter plant roots, extending the clone. Other plants produce bulbils attached to the flowering stem; they are much less costly to produce than seeds but neither travel as far nor are as resistant. Many plants adopt the bulbil strategy; they include the nodding saxifrage (*S. cernua*), the grained saxifrage (*S. foliosa*), and the viviparous knotweed (*Polygonum viviparum*).

High Polar Deserts

Once the climate becomes so severe that the tundra's community structure breaks down, we find plants hidden in the nooks and crannies between the loose rock that passes for soil in the High Arctic lands. In these regions the plant cover is always less than 5% of the ground surface (except in a few oases fed by melting snow); as a result, competition plays only a small role in plant survival. In these High Arctic regions, only the physical microenvironment determines whether a seed germinates and whether a seedling survives. These environments are so cold and dry that only a few dozen plant species have managed to colonize here. This is truly the realm of belly flowers (unless you're resting on your belly, you can hardly see them, let alone truly appreciate them).

In these arctic deserts, with degree-days less than 180, only a few body plans allow for plant survival. One of the most common is the miniature-rosette body plan. It is so popular that it has been adopted by four different families: the saxifrages, the pinks, the mustards, and the poppies. Another successful one is the prostrate-grass body plan. In the High Arctic, grasses seem to be particularly suited to sandy soils that have a somewhat deeper active layer (the layer that thaws out in summer) than the pebbly soils normally found. Other places dominated by grasses include salty soils and soils that undergo erosion during snow melt. In these regions, edaphic (soil) factors play a big role in plant survival.

The Canadian High Arctic covers areas where the soil is derived from limestone, sandstone, shale, or granitic parent material. These four types of materials favor different types of plants. In addition, any promontory or slightly elevated object, be it bone or boulder, will attract animals, especially birds, who will proceed to enrich the soil around the object. Growing beside such an object affords not only protection from the wind but also moisture and nutrients. Some plants grow in such fertilized areas and almost nowhere else. It is perhaps surprising that even with this level of heterogeneity, two of the most successful and ubiquitous plants in the cold deserts of the High Arctic are the arctic poppy, a miniature-rosette species, and the purple saxifrage (*Saxifraga oppositifolia*). They grow in almost every kind of soil and almost every moisture regime, from late-melting snowbanks to uplands that get only a few centimeters of snow in the winter. The purple saxifrage is exceptional among the arctic desert species in that it is a semiwoody plant and does not fit neatly into either of the body plans mentioned above. Instead, it is densely matted with condensed, crowded, somewhat woody stems. It can be found growing almost everywhere in the High Arctic, although it prefers calcareous substrates and is not found in very acid soils. It can be found even in deep snowbanks that melt in late July, thus giving the plant very little time to flower and set seed. Its ubiquitousness in the High Arctic is contrary to the experience of many an alpine enthusiast. When cultivated, this species' seedlings are slow to get established, and once established cannot withstand drought, especially in the spring. In cultivation it needs a humus-rich soil, while in the High Arctic it grows in bare mineral soil. These differences can be attributed to the very low temperature and high relative humidity throughout the High Arctic, factors that decrease the transpiration load on these plants. Their susceptibility to drying out is also expressed in nature at the southern limit of their distribution. In the Rocky Mountains this limit occurs in northwestern Wyoming, where these plants grow only in moist though well-drained sites.

Snowflush Communities

Snowflushes are deep, protected snowbanks that melt in late summer; they are unique environments that tax the abilities of even the speediest plants. In the High Arctic the only vascular plants to be found in late-melting snowbanks are the purple saxifrage, the arctic poppy, and the nodding saxifrage. Sometimes, along with these, we find members of the pink family (Caryophyllaceae) such as the boreal sandwort (*Minuartia rubella*) and a chickweed, *Stellaria crassipes*. These plants probably have the lowest requirements in terms of degree-days of any plants on this planet, but I am not aware of any detailed measurements. Snowflush communities also exist on every mountain range that is tall enough and receives enough moisture, and we will discuss these in chapter 10. For now it is enough to realize that snowflush communities of different places have different controlling factors and different compositions. For example in the European Alps, a common buttercup, the glacier crowfoot (*Ranunculus glacialis*), has been known to survive up to 33 months under a continuous cover of snow, which are much snowier conditions than High Arctic species ever experience. Under heavy snow, vegetation can be encased in ice, and in the darkness it can be exposed to prolonged anoxic periods. Therefore, not all the deleterious effects suffered by plants in snowflushes can be attributed to just low temperatures or low light levels.

Sunlight is necessary for the growth of every plant, though sometimes not directly, as is the case with parasitic plants such as dodder (*Cuscuta gronovii*). If sunlight is weak, as it is on the forest floor or under a blanket of snow, plants must evolve more efficient methods of harvesting light if they are to make these places their home. Many plants have found ways to do this, as is evident from the large number of tropical forest floor plants that have found adequate accommodations in kitchens, foyers, and bathrooms around the world. In this new habitat the availability of light is of much less concern than the vagaries of irrigation. One of the most common adaptations to low light intensities is the creation of a reflective underlayer of the leaf so that there is a second chance to capture light photons and use their energy in photosynthesis. This modification seems to have no familial boundaries, and it is found in Moses-in-the-cradle (*Tradescantia spathacea*), so popular in south Florida homes, as well as in many begonias. Vascular plants such as these, however, seem to have genetic limitations that preclude the adaptation to extreme shade that is found in algae and bacteria. Extreme shade requires new pigments (and the concomitant genes) that are able to interact with unused sections of the electromagnetic spectrum.

In the Arctic, a few studies have shown that certain plants, including mountain sorrel (*Oxyria digyna*), have temperature and light intensity optima that are lower than those of alpine plants from more southerly climes. Mountain sorrel is a ubiquitous snowflush community member, and this may explain why older leaves of this plant are often dark red or even plum colored. The dark color might help the plant melt the snow around its leaves, thus increasing light intensity and also the ability of the leaf to exchange gases with the atmosphere. This greenhouse-under-snow may be a common adaptation of snowflush plants. Other snowflush plants, such as buttercups, have deep reflective layers made of starch granules that make their leaves shiny green and might work to enhance the capture of light. When the snows melt very late, be it in the polar regions or on alpine slopes, no vascular plant can survive, and the snowflush realm returns to the care of its antediluvian inhabitants, the mosses.

Water in the Arctic

In the arctic tundra water is normally in plentiful supply. Permafrost does not allow water to percolate into the soil, and since the moisture-holding capacity of cold air is very limited, water simply accumulates in even the slightest depression. Except for slight frost-induced ridges, the tundra is a soggy place. However, High Arctic deserts are normally found on sloping uplands where summer precipitation is low even by arctic standards (20 to 60 mm a year), and where snow cover is minimal (less than 20 cm). This amount of precipitation is barely enough to compensate for yearly evaporation (50 to 90 mm), and no water collects on the surface of the land. In most places there is still permafrost, but the active layer can be as deep as 80 cm by early August. Spring melt from areas of snow accumulation and that from glaciers are the only reliable sources of water. Many times the water coming from these sources flows below the surface of the soil, constrained and guided by the layer of permafrost just below it, until it finds a slope steep enough for it to burst its confines, emerge at the surface, and trickle down the slope. These seeps manifest themselves as lush moss-covered areas that may contain the only arctic willows of these polar barrens.

Other areas, not blessed by either surface or subsurface flows, are bone dry by early summer. Plants growing under such conditions must obtain water early, when the meager snow that covers the ground in spring melts. At this time the ground is very cold, and these plants must have special adaptations to obtain all the water they need for the following summer's growing season. Obviously, the earlier these plants can start metabolic activity the better, and again purple saxifrage is one of the best-adapted plants in this respect, being the earliest flowering plant in many parts of the Arctic.

Glaciation at Both Poles

If the lack of crustose and foliose lichens is any guide, the High Arctic environment that arctic plants today call home is relatively new. The areas in which these plants are growing now may have been covered by glaciers or permanent snow fields as recently as the Little Ice Age that lasted from about 430 to 130 B.P. If this is the case, there may be other species just as competent to stand the rigors of these arctic deserts, but that simply move more slowly when reconquering territory. This is an important point whenever we discuss the limits to life on Earth. The reason that there are several dozen species of vascular plants (plus a dozen or so mosses) able to stand the rigors of the High Arctic deserts, but only two vascular plants that even venture onto the 4% of Antarctica that is ice free, is not so much that Antarctica is that much harsher, but that it is that much more isolated and has been so for a very long time.

Forty million years ago Australia completely separated from Antarctica. The antarctic continent was then left to drift slowly into inexorable, cold oblivion over the next 20 million years. Twenty million years ago it arrived at its present location, with a shield of snow and ice that would eventually hide most traces of its living past. The plants and animals that had evolved in Antarctica did not stand a chance (except for the penguins) as the ice sheet expanded and eventually engulfed the whole continent during the glaciations of the past 2 million years.

The southern beeches (*Nothofagus*), other flowering trees, and conifers (*Araucarioxylon*, *Araucaria*, and others) flourished in the South Shetland Islands (the last bastion of forests in Antarctica) 16 million years ago in what must have been temperate forests very similar to those of Patagonia today. They did not have the genetic equipment to adapt to a very low heat sum and extremely low winter temperatures. This can be seen today in both Patagonia and New Zealand, where the southern beeches dominate the alpine timberlines. These trees are limited by their ability to frost-harden their foliage during the short growing season and may succumb to winter desiccation. This explains the generally lower altitudinal limit of trees in the Southern versus the Northern Hemisphere at equivalent latitudes. That some North American timberline species are better adapted to this antipodal subalpine environment than the native Southern Hemisphere species has been shown by experimental planting of lodgepole pine (*Pinus contorta*) in New Zealand. At some sites, the lodgepole pine is advancing up the mountains past the native beech timberline, whereas in its native North America the lodgepole pine is seldom a major component of the upper timberlines except in fire-prone areas.

For the kingdom of plants to try to reclaim Antarctica, it would have to summon its most adventurous knights. It is no coincidence that the two vascular plants to recolonize the Antarctic Peninsula belong to the pink and grass families; these pages are full of examples of their conquering a variety of harsh environments from hot deserts to salty shores to arctic deserts. This may be expected from the grass family which, with over 10,000 species, is, after all, one of the largest families of flowering plants. The pink family, with only about 2400 species, is more surprising. What is even more surprising is that some genera in these two families can range from the northern shore of Greenland through the Irano-Turanian steppe to the deserts around the Persian Gulf. One of these is the genus *Silene*: the catchflies or campions. For one genus to span such a wide range of environmental variables it must have great genetic plasticity. The populations of the arctic species and those of the desert species must have completely different temperature optima; this probably requires small mutations affecting almost all the enzyme systems. They also must have completely different methods of dealing with water stress, since the arctic plants can simply close their stomata to conserve water, but if the desert plants do this, they are in danger of suffering heat damage.

Insects in the Arctic

Insects are crucial to the survival of many of the plants we have discussed thus far; even the purple saxifrage relies extensively on the arctic bumblebees for pollination. Many arctic flowers are able to pollinate themselves at least to some extent, a condition known as selfing. A high selfing rate helps a plant spread faster than if it had to wait for the concomitant spread of pollinators. The arctic poppy is a very good selfer; but the arctic dryad only manages to ripen 40% of its usual seed crop without the aid of pollinating flies. The purple saxifrage is a poor selfer; in this plant, seed yield without the aid of bumblebees is only about 6% of that with the aid of bumblebees, and for the moss campion (*Silene acaulis*) the figure is closer to 1%. This tight interdependence between bumblebees and bumblebee flowers slows the spread of both into newly exposed land. The bumblebees cannot move into these lands without the refueling stations provided by the flowers, but the flowers cannot spread into these areas if there are no resident bumblebees. Even for a very tough species like the moss campion, this interdependence slows its conquering of newly exposed land and suggests that the ultimate possible diversity to be found in the High Arctic may only be achieved if time and climate allow it. Other plants such as the grasses and willows are pollinated by the ubiquitous agency of wind. Wind has none of the single-minded purpose of insects and cannot be finely honed by the force of coevolution to provide unerring delivery of pollen, but it is powerful and omnipresent. Wind-pollinated plants waste no effort making attractive petals or sepals; instead they spend their resources and energy on pollen, and lots of it.

Flies make up the bulk of insect biomass in the Arctic and also the majority of insect species there. In the Arctic, flies are crucial not only as pollinators of many plants, but also as critical members of the soil fauna and the aquatic fauna. Flies are part of the insectivorous food webs of the Arctic, in which the top predators are wading and perching birds. One of the most diversified groups of flies in the Arctic is that of the crane flies (Order Tipulicida). One of the crane flies, *Pedicia hannai*, preys on enchytraeid worms (Class Annelida: Order Oligochaeta), which are plentiful in arctic soils; this is a miniature version of the mole-earthworm relationship of warmer latitudes. Crane flies are the main food for the Lapland bunting and its nestlings at the height of summer. These flies are also important members of the detritus-based community, and they are partly responsible for whatever nutrient cycling and soil formation occurs in the polar regions. Another group of Arctic flies, the chironomid midges, is the cornerstone of aquatic ecosystems that lead to freshwater fish and eventually to humans. In the High Arctic, the Anthomyiidae is among the most evolutionarily energetic families of flies. This family's special talents seem to be spurring a new bout of evolutionary expansion with over 130 catalogued species, six of which reach the High Arctic. The members of this family feed on decaying and living plant tissues during the larval stage; as adults, they feed on nectar and pollen. Only one other family of flies, the Muscidae, to which the common housefly belongs, has a greater number of species in the High Arctic. The insect residents of the Arctic and Antarctic outnumber the vertebrate residents by several fold. Part of this success can be attributed to their adaptations to cold.

Some of the most freeze-tolerant arctic insects have received little study, since many times it can be difficult to collect the overwintering stages (normally the larvae) in the wild. For instance, one of the most ubiquitous and distinctive insect larvae of the Canadian High Arctic and Greenland is the woolly bear caterpillar (*Gynaephora groenlandica*), which is the larva of a lymantriid moth. This is one of the first insects to become active in the arctic summer, yet it disappears by early July when most other insects are at their most frenetic. To find arctic insects, a collector must know not only when, but also where to look. Many times the easily collected forms are adults in their mating flights. In the High Arctic, the paucity of plant and animal matter suitable for adult insects has forced many species to forgo a lengthy stay in the adult phase; often adults are ephemeral creatures that do not even have mouthparts and therefore do not feed. The adults depend on whatever reserves they stored as juveniles; when these reserves are exhausted, whether they have successfully mated or not, they die. To find the cold-tolerant stages of these insects, we must seek out the cryptic larvae. The environments these larvae occupy are diverse, but they all provide a certain amount of protection to the inhabitants.

Some insects, especially some flies and wasps, have found comfortable, though temporary, accommodations in the bodies of other organisms. These are the parasitoids. A parasitoid is a most insidious sort of predator, depending on its prey not only for food, but also for shelter. A parasitoid has little in common with a parasite; the former exploits and sooner or later inevitably kills; the latter, if it is successful, exploits but seldom kills. A case in point is the bristle fly (family Tachinidae) that is a parasitoid of the woolly bear caterpillar. Flies belonging to the Tachinidae resemble a cross between a housefly and a hedgehog. The Tachinidae is the second largest family of flies in North America and consists mainly of parasitoids of other insects. Female bristle flies lay eggs directly on the host or on the food plants of the host. In the latter case the eggs hatch in the host's digestive tract, whereupon they start feeding on its nonvital tissues. Two-thirds of woolly bear caterpillars have some sort of parasitoid, either a bristle fly or an ichneumon wasp. In the case of the bristle fly, one to sixteen fly larvae are found within each parasitized caterpillar. The fly larvae develop over several years, feeding on the caterpillar's nonvital organs until it pupates. They pupate inside their host's pupa and emerge as adult flies in the spring, in the process killing the host's pupa. The ever-curious polar explorer James Ross made some observations on the woolly bear caterpillar:

> About thirty caterpillars were put into a box in the middle of September, and after being exposed to the severe winter temperatures of the next three months, they were brought into a warm cabin; where in less than two hours every one of them returned to life, and continued for a whole day walking about; they were again exposed to the air at a temperature of 40 below zero, and became immediately frozen; in this state they remained a week, and on being brought again into the cabin, only twenty three came to life; these were at the end of four hours put out once more into the air and again hard frozen; after another week they were brought in, when only eleven were restored to life; a fourth time they were exposed to the winter temperature, and only two returned to life

> on being brought into the cabin; these two survived the winter, and in May an imperfect Laria [Gynaephora] was produced from one, and six flies from the other. (Ross, 1835)

The extreme resilience of the caterpillars is of note, but equally remarkable is the resilience of the six fly larvae to this temperature roller-coaster ride. We will delve at depth into winter cold tolerance in the next chapter.

Winter-Active Residents of the Arctic

In the polar regions, both mammals and birds have taken insulation to its ultimate development. If the internal fires can be kept stoked and the core body temperature does not drop, animals such as the arctic hare do not seek shelter during blizzards in which temperatures can plummet to -40°C. Of course, in order to stand such cold, animals must be well fed. During times of heavy snow, when they cannot dig down deep enough to reach their winter food supply, there might be mass starvation and death from cold. This has happened several times in recorded history in the North American Arctic. There have been dieoffs of caribou in Greenland and of muskoxen and arctic hares in the Canadian Arctic Islands. The death of whole migrating flocks of birds has also been recorded.

Of the many birds that breed in the polar regions, the vast majority migrate south as winter approaches. Migration is a very successful cold-avoidance mechanism, but what is of interest here is cold tolerance. Only very few bird species can claim the title of arctic or antarctic winter residents. In the Arctic, phytophagous (herbivorous) birds depend on the shrubby plants that reach the snow surface in winter, and the availability of these plant species defines the birds' geographic distribution.

Compared to beak and hind legs, forelegs are much more easily adapted to the task of digging through snow; this enhanced digging morphology allows mammals to exploit the subnival environment more effectively than birds. A couple of mammals have evolved to exploit the subnival world of the Arctic. These are two microtine rodents: the brown lemming (*Lemmus sibiricus*) and the collared lemming (*Dicrostonyx* spp.). These rodents have different preferences in terms of microtopography and food plants. The brown lemming is a Low Arctic species that seems to be expanding its range northward; it prefers wetter cottongrass tundra where watersedges (*Carex* spp.) and cottongrasses predominate. Here it feeds on the tender shoots of cottongrasses as well as on arctic willow, which also thrives in these areas. In winter it eats mostly mosses and is one of the few animals that can survive on this diet. On the other hand, the collared lemming reaches the High Arctic and the Ungava Peninsula, and even into northern Greenland; it prefers drier habitats where arctic willow and arctic dryad, along with purple saxifrage and louseworts, thrive. In good years lemmings may have up to 3 litters of 5 to 10 young each. This fecundity translates into plenty of prey for the land carnivores that share these northern lands. The arctic fox and the snowy owl are the main predators of the lemmings, and in good years the foxes may raise 10 or 11 pups in a litter, while the snowy owl may raise 3 to 4 chicks. In bad years, when the lemming population falls, the predators might not be able to raise a single offspring and might be forced to abandon them.

Two other avian winter residents of the Arctic are the rock ptarmigan (*Lagopus mutus*) and its cousin the willow ptarmigan (*Lagopus lagopus*). Together they cover the entire arctic region from the arctic coast of Alaska and that of Siberia to the Canadian Arctic Islands, Greenland, and arctic Europe. Their adaptations include snowshoe feet (as in the snowy owl). In fact, the genus name comes from *lagos*, hare, and *pous*, feet. Their feet are always feathered on the top, and feathers grow on the bottom as winter approaches. The feathers allow these ptarmigans to increase the surface area of their footprints by 400%, and experiments have shown that birds with their feet plucked of feathers sink into the snow 50% deeper than normal. The winter diet of these galliformes includes winter buds of willow and birch and twigs of these two species. Whether willow is preferred over birch depends on the availability of both. The willow ptarmigan is more restricted to willow scrub than the rock ptarmigan, which prefers more open, drier tundra. In the Alaskan Arctic, the winter diet of the willow ptarmigans consists almost entirely of feltleaf willow twigs and buds; in other areas, other species of willow predominate in its diet. The rock ptarmigan, on the other hand, feeds mainly on the winter buds of birch, but its diet is not restricted to them. In winter, both these birds will excavate into the snow to reach their food supply and also to insulate themselves from the bitter cold. They enter the snow by diving into it from the air in order to leave no tracks that would give away their position. North of Melville Island and Bathurst Island (in the Canadian Arctic), in northern Greenland, and in Svalbard (a Norwegian archipelago), the only bird that overwinters is the rock ptarmigan. This, however, is the limit of its distribution, and in certain years most of the birds move south several hundred kilometers to milder climates. In fact, most populations of ptarmigans migrate south short distances, and females migrate farther than males. However, their wintering ranges are still well within the Arctic.

The arctic redpoll is the High Arctic variety of the common redpoll (*Carduelis flammea*) and is the only passerine to overwinter in the Arctic. Its winter diet consists of birch seeds, and its summer diet consists of insects, mainly crane flies and midges, supplemented by a few seeds and berries. It spends a substantial portion of its time under the snow, like ptarmigans do. Besides its behavioral adaptations, the redpoll, unlike many other song birds, also has a crop—a morphological adaptation to its arctic environment. This extension of the esophagus is used to store some of the food the bird gathers while foraging, so it can be digested at leisure under the protection of snow. For nesting sites, instead of digging its own burrows, it uses vacant lemming burrows, which it lines with ptarmigan feathers and the down of willow and of cottongrasses. Its use of lemming burrows may mean that arctic redpoll populations may be more limited by lack of appropriate nesting sites than by lack of food.

Large Animals

Because of their small size, the previously mentioned herbivores make their winter homes under the snow and thus avoid the worst of the winter weather. Weighing in at 5 kg, the arctic hare (*Lepus arcticus*) is the largest of the hares and is one of the hardiest mammals of the High Arctic. It can be found on all the High Arctic Islands except for Prince Patrick Island. It also ranges east to Baffin Island and south to the arctic coast of North America. The population in the High Arctic is distinct in its coloration and behavior; it is all-white all year round (except for black ear tips) and tends to stand on its hind feet much more often than hares farther south. This behavior seems curious to us because we are so used to being the only erect mammals, but it makes perfect sense for the hares, since the tallest visual obstacles they have to contend with are other hares! Unlike the smaller mammals already discussed, the arctic hare does not seek subnival shelter. Over its range it prefers arctic semideserts where various species of grasses predominate. These environments are normally found on highlands where little snow accumulates, and the hares in fact retreat to windblown ridges to spend the winter, since these are the places where their winter food, arctic willow, can be found. In these wintering grounds the hares might be exposed to average monthly temperatures of -38°C, but as long as there is enough exposed willow, the animals can survive.

There are several subspecies of caribou and reindeer (*Rangifer tarandus*) in the north polar regions. Unlike many creatures that migrated west to east across the Bering land bridge, the ancestral caribou, along with the ancestral horse and camel, migrated from North America into Eurasia. The smallest of the current set of subspecies is the Peary's caribou (*Rangifer tarandus pearyi*). This subspecies of caribou evolved under the harsh climatic conditions found on the Canadian Arctic Archipelago. The population dynamics of this caribou are closely tied to the winter

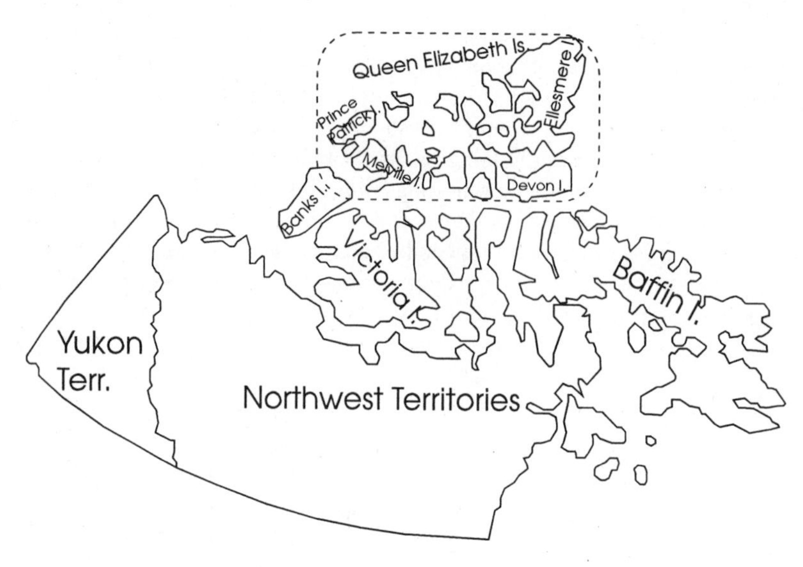

The Canadian Arctic Archipelago, including the Queen Elizabeth Islands.

availability of food, which in turn depends on the depth of the winter snow and ice cover.

As is the case with other deer, caribou population numbers can vary widely, and this inherent fluctuation is exacerbated by the harshness of the climate. A couple of decades ago, a series of severe winters decimated the caribou population in the western Queen Elizabeth Islands. The caribou population in the southern part of their range dropped to 32% of its former value after the winter of 1973-74; few if any of the calves born that year survived the winter. A late summer ice storm can also decimate a population, since the animals cannot break through groundfast ice; the same thing can happen during spring melt if a cold spell refreezes the melting snow.

The Peary's caribou migratory movements, unlike those of other caribou subspecies, vary depending on local conditions, but few populations ever leave the Canadian Arctic Archipelago. Even though Peary's caribou share their range with muskoxen, the diets of these two animal hardly overlap. In the harsher areas of their range, caribou feed mostly on graminoids (mainly the woodrush *Luzula*) and mosses, with a smaller proportion of lichens rounding out the diet. Even though in the popular mind caribou are the preeminent lichen feeders, in the High Arctic the types of lichens available are mainly crustose ones, which are closely appressed to the rocks on which they grow and mostly unavailable to grazers like the caribou. The low primary productivity and low standing crop of lichens also contribute to their absence from the diet of most vertebrates in the High Arctic.

The other large herbivore in the High Arctic is the muskox (*Ovibos moschatus*). This is an imposing animal; even though the larger males stand only 1.5 m at the shoulder, the long-flowing guard hair projects a much more massive appearance. Muskoxen are about the best cold-adapted mammals on the planet, but even they are at the mercy of the polar winter. Like so many present-day North American mammals, the muskox is an immigrant. In the Pleistocene there were six different species of muskox in the Siberian tundra and Siberian taiga. During one of the more recent glaciations, the species we now know as the muskox occupied the unglaciated parts of northeastern Siberia and northwestern Alaska, including the land now submerged by the Bering strait.

The above-mentioned unglaciated land has been christened Beringia, and it played a very important part in the biological history of northern Asia and North America, since it served as the largest ice-free refuge for plants and animals throughout the glacial periods. What Beringia looked like at the height of the glaciations is a hotly contested subject. One camp of researchers thinks that it looked like a polar steppe and that the paleontological evidence supports this theory. This camp envisions an African-savannah-like environment with vast areas of pastures supporting large grazers like mammoths, woolly rhinoceros, and of course, muskoxen. The other camp points out that the large concentration of bones sometimes found does not necessarily mean large total populations over Beringia, but could be explained by the existence of populations localized in biological oases as those seen in the High Arctic at the present time. This camp's

The muskox (*Ovibos moschatus*).

argument is based on the absence of large amounts of fossil pollen of grasses and other steppe plant species, which would be expected if the steppe theory were true. I suspect that the answer may lie somewhere in between these two positions.

Once the glaciers started to melt, muskoxen in North America dispersed throughout the Canadian arctic coast and the Canadian Arctic Archipelago, eventually reaching the eastern coast of Greenland. The Siberian populations did not fare well, however, and a combination of hunting pressures exerted by neolithic hunters and the warming of the climate drove all the Siberian muskoxen to extinction. Even on the arctic coast of Alaska, the last remaining modern muskoxen became extinct by the middle of the nineteenth century due to relentless hunting pressures. Elsewhere, when these pressures were eased, population numbers started to rebound, and several attempts were made to reintroduce the animal to parts of its former range. In 1935, animals from the largest population, the one in northeastern Greenland, were taken to Nunivak Island in the Bering sea, where they thrived. The population in Nunivak has become the progenitor of several populations reintroduced to the North Slope of Alaska's Brooks Range and introduced to Barter Island, a tiny island off the arctic coast of Alaska. At present there are significant populations along the Canning and Jago river drainages on the North Slope. Some of the animals from Nunivak have also been reintroduced to Siberia and Wrangel Island, where there are now several thriving populations.

Not all introductions have been successful, though. In the early part of the twentieth century, the introduction of arctic mammals to distant lands seems to have been all the rage. Some introductions were thought through reasonably well (given the state of knowledge at the time about arctic mammals' requirements). Some introductions had a chance of success, especially those to parts of an ancestral range. However, most introductions were not to ancestral ranges. In the case of the muskox, an attempted introduction to Svalbard in 1929 seemed to succeed for a while, but by the early 1950's the population was in decline, and it became extinct in 1959. Another case was the attempted introduction of the muskox to Iceland on three separate occasions. In the first attempt in 1905, seven calves from northern Greenland were released in a lush valley in the southwest of Iceland; they soon died due to the unsuitability of the habitat. Again in 1929 seven animals were introduced to Iceland; all of them died of disease within three years. A third attempt in 1930 met with a similar fate. From this and similar failures, it becomes obvious that for an introduction to have a chance of success, first all the information on the requirements of the particular species must be gathered.

It can be said that the muskox, compared to the other arctic mammals that have been discussed, is the one that has taken morphological adaptation close to its logical extreme. It has maximized its thermal time constant by increasing its size, decreasing its surface area (by decreasing the length of its extremities), and increasing the efficiency of its insulation. However, a large thermal time constant provides no advantage unless the organism can maintain its internal heat production above a certain minimal level.

In the High Arctic winter, food is scarce and relief from the cold can be months away. The trick to survival is to minimize the level of activity in order to minimize energy expenditure and stretch food resources and internal energy stores as much as possible. Under such circumstances it makes sense to conserve energy and move only in search of new graze. In many places in the Arctic, muskoxen spend about as much time standing still as grazing. Even with these adaptations, they are vulnerable to the vagaries of arctic weather. In the late winter and early spring, when they are the weakest, the rising water table forces them to switch from their preferred browse of cottongrasses to willows and grasses on higher slopes. Late snows at this time can cover these higher slopes and cause starvation.

Antarctic Residents

Just like in the Arctic, there are many animals that raise their young in the Antarctic and migrate to warmer climates as winter approaches. However, we are interested only in those organisms that have found a way to survive as year-round residents of the Antarctic. There are fewer year-round residents of the Antarctic than of the Arctic. There are several reasons for this. I have already mentioned Antarctica's complete encasement in ice during the ice ages that have occurred over the past 2 million years. This entombment sealed the fate of almost all the terrestrial inhabitants of this continent. (It took until 1979 for expeditions to finally find fossilized traces of mammals in the Antarctic peninsula.) A second factor responsible for the lack of residents is Antarctica's isolation, which has precluded the return of land mammals after they were extirpated by the cold and ice sometime in the past 14 million years.

The only land animals that currently call Antarctica home are two species of insects, several species of mites, several species of springtails, and two species of copepods. These arthropods make their homes in tufts of grass or in mosses, or in the few lakes that are not permanently covered with ice. Of the two species of insects, one is found only on Signy Island and the other, the chironomid midge *Belgica antarctica*, is found south to 65°S latitude on the Antarctic Peninsula. These animals are sustained, in terms of both food and shelter, by the mosses and the two vascular plants that make this part of the Antarctic Peninsula their home. One of these vascular plants is the antarctic hairgrass *Deschampsia antarctica*, and the other, *Colobanthus quitensis*, has been called a pearlwort, among other names. Neither of these two vascular plants is completely adapted to life in the freezer that is Antarctica, but they manage mainly by opportunism. In the low light intensities found there, the grass is seldom operating at its maximum photosynthesizing efficiency, while the pearlwort can sometimes reach peak efficiency. From this consideration alone, we would expect the pearlwort to be the dominant plant over the grass, but the opposite is the case. Other factors intervene, including the difficulty of germinating and surviving as a seedling in such an inhospitable place. In contrast with the vascular flora, the moss flora consists of over a dozen species and the lichen flora of several dozen, the opposite of what is found in the High Arctic. This may be due to the higher relative humidity found in the maritime Antarctic as compared to the High Arctic.

No vertebrate in Antarctica can be called truly terrestrial, and all directly or indirectly derive their sustenance from the sea. Even the littoral biome is almost totally absent from the ice-free coasts of Antarctica. Just like in the Arctic, there are marine organisms that frequent this zone at least in summer; they retreat into deeper water in winter as the shallows are first exposed to scouring ice and then completely replaced by a layer of ice. Only two birds seem to take advantage of the littoral zone in summer. These are the southern black-backed gull (*Laurus dominicanus*) and the American sheathbill (*Chionis alba*).

The southern black-backed gull is closely related to the northern black-backed gull and like its relative feeds extensively on mollusks, particularly Antarctic limpet (*Nacella concinna*), which it swallows whole, regurgitating the almost empty shell. These birds never stray far from land, and even their winter migration takes them to the outer Antarctic islands where the nearshore ice is absent or not as extensive, and they can hunt their traditional prey.

The many sightings around the continent's coast suggest that the birds are expanding their nesting range. They nest on the Antarctic Peninsula to 68°S latitude and on the Antarctic islands. Their population expansion is probably linked to their propensity to feed at the refuse dumps of the permanently manned Antarctic stations. The creation of permanent human habitations has produced a year-round food source that is being exploited by several of the Antarctic residents. Besides the gulls, snow petrels (*Pagodroma nivea*) and sheathbills also frequent these dumps. Especially for the gulls and sheathbills, dumps allow the birds to remain near their nesting grounds throughout the winter.

While the black-backed gull looks like a gull, the sheathbill reminds one of a corpulent domestic pigeon both in appearance and demeanor; it even walks with the characteristic head-bobbing gate of pigeons. It is, of course, not related to pigeons, and the two species of sheathbill form their own family (Chionididae) structurally poised between the shorebirds (such as the lapwings) and the gulls. The genus *Chionis* is thought of as primitive, possibly similar to the ancestor of those other more advanced groups. The sheathbill's terrestrial nature can be seen in its feet; the three front toes have only rudimentary webbing between them, and the fourth is strongly opposed, suggesting a bird built for beach combing and not for swimming. Its behavior also suggests a terrestrial bird; it seldom enters the water even though it is a capable swimmer.

The sheathbill is not a fussy eater; it will not turn down anything of nutritive value. It has very undiscriminating tastes and will eat insects, seaweed, seal placentas, and even seal and penguin feces. As this smorgasbord suggests, the sheathbill's nest is neither tidy nor very hygienic, and if this bird lived in warmer climes, its habits would likely attract unwanted nest visitors. The sheathbill is quite cunning in its utilization of resources. It is cunning enough to steal the meal from a penguin chick's mouth, and on rare occasions it may kill a young chick. More often, however, it acts as a terrestrial scavenger feeding on the corpses of recently dead animals. It also feeds on the leavings of other animals, including the limpet shells discarded by the black-backed gull.

The emperor penguin (*Aptenodytes forsteri*) is possibly the most remarkable inhabitant of Antarctica. It is one of only two species in its genus, the other being the king penguin (*Aptenodytes patagonicus*), a slightly smaller bird but with a story almost as remarkable, since its breeding cycle spans a three-year period. The emperors are perhaps the best recognized of the penguin clan, with their long beaks, tall stature (for a penguin), and orange-and-red neck blotches. Perhaps because of their longer necks and bills, they look more distinguished than other penguins.

Most emperors nest on ice, never setting foot on solid earth, in a sense making them as aquatic as a bird can get. Females start arriving at the nesting grounds in early March, just as the cold winds are starting to change the character of the Antarctic seas. The males arrive a little later, and in a cacophony of introductions and renewed liaisons, these animals find mates while the sea congeals under the force of cold. Once mating is completed, the female lays one large egg that may weigh as much as one third of her body weight. The male eagerly wraps his feet around the egg, and with skillful head and beak movements he manages to place the egg on top of his feet. He then covers it with a flap of feathered skin protruding from his lower belly. The female, having completed her task, turns around and marches toward the sea, to feed and replenish her strength. The males are left alone to endure the worst of the Antarctic winter with only their fat reserves to sustain them. As the females enter the sea, it steams and rolls as if it were a freshly made vat of molasses. The sea is now a viscous liquid crystal giving up its heat in wisps of heavy vapor, and by the time the females return to the colony, the sea will again have frozen into an unbroken sheet of ice.

The cold has driven emperors into a much more social behavior than that of other penguins. Whereas other penguins bicker constantly over territorial boundaries, the emperors conserve warmth by pressing against their fellows. The birds at the periphery are driven by the unrelenting cold winds to seek shelter at the center of the mass of bodies, thus exposing others who will soon join the cyclical march. By the time the chicks are ready to hatch at the end of three months, the males have lost 30% of their body weight. If a female doesn't reach the colony within a few days of the chick's hatching, the male will be forced to abandon it in order to save his own life. Even if the female is not present when the chick hatches, the male feeds it with an oily exudation of its crop, the last resort of an extremely dedicated parent. Once the females start arriving at the colony in late May,

The emperor penguin (*Aptenodytes forsteri*) with chick.

the two sexes divide the tasks of hunting and brooding until the chicks are large enough, by three months of age, to gather in relatively unsupervised crèches.

By laying at the start of the austral winter, and brooding before the migrant skuas and giant fulmars arrive, these penguins have completely sidestepped predation as a cause of nestling mortality. Their biggest enemy is not cold per se, but the effect the cold has on the expanse of sea ice females have to cross on their return trip to the colony. In years when the sea ice is unbroken for many miles away from the colonies, chick mortality can be high. This especially applies to the more southerly colonies like Cape Crozier, where the females may have to walk over 100 km to reach the colony from the sea.

Antarctic Lakes

There are some remarkable aquatic habitats in the little bit of Antarctica that is free from the perpetual clutches of ice. These are the lakes of Victoria Land, at about 150°W longitude. About a dozen of these lakes have been found in several extremely dry valleys that lie in the precipitation shadow of the Transantarctic and Admiralty mountains. These two mountain chains encircle the valleys and in the process produce one of the most inhospitable environments on Earth. Yearly precipitation is less than 100 mm, and the howling winds that are funneled down these valleys remove what little snow falls. Such a small amount of precipitation means that the lakes that do exist in these valleys are arheic; they have no watersheds to speak of and are instead fed by the intermittent streams born in nearby glaciers. These streams thaw only in certain years, and the lakes themselves are never ice-free. However, most years the shore ice melts, and a moat that allows influx of water and nutrients forms around the lakes.

Since very few of these lakes are connected, each is in effect its own little world. Geochemically, each lake has evolved down its own distinct path that depends on the morphology of the valley, the composition of the water flowing in, and the substrate the lake overlies. All of the lakes are fresh near the surface, but a few are hypersaline at depth.

A hallmark of these lakes is their ice cover. If these lakes were ice free (an impossibility under the current climate), the mixing induced by wind would create much more hospitable conditions for their planktonic inhabitants. Without wind, phytoplankton must float or swim to stay in the lighted reaches of the water column; otherwise the little plants will sink into an abyss not measured in fathoms but in units of light, an abyss in which they will pass from the realm of poor light to the realm of absolute darkness. Such a rain of organisms takes valuable nutrients out of the photic zone and into the little-known realm of decomposers at the deepest recesses of these lakes.

Another unavoidable loss from the system occurs every winter as the ice thickens and steals fresh water from the lake, leaving a saltier, denser layer of water that sinks until its density is in equilibrium with the surrounding water. Several other processes remove essential nutrients from the water column. These processes make the lakes so nutrient-poor (oligotrophic) that they may be compared to alpine lakes with similarly low biological diversity.

Limnologists often speak of the compensation point where light becomes so dim that no plant can photosynthesize enough to compensate for its own respiratory needs. Oligotrophic lakes tend to have deep compensation points, unless the water is turbid because of suspended sediments like rock flour or clays. This turbidity tends to decrease light penetration and bring the compensation point closer to the surface; this sometimes happens in alpine lakes fed by glacier streams. But the Antarctic lakes receive so little fresh-water input and wind mixing that the small amounts of suspended sediments hardly affect the compensation point. In cold waters (as in most of the Antarctic lakes), respiration is inhibited more than photosynthesis, and the compensation point is found at lower light intensities, and therefore greater depths, than if the waters were warmer.

Lake Vanda is the deepest (over 70 m) of these Antarctic lakes, and its waters are also the warmest, reaching a temperature of 25°C at the bottom. This heating effect is due to the lake's very oligotrophic nature (the very low organic content lets 1 to 2% of the surface sunlight penetrate all the way to the bottom) and to its chemical stratification. At a depth of between 50 and 60 m, the salinity increases abruptly. This salinity increase traps the heat generated by the absorption of sunlight, but it has other effects as well. Below a depth of 58 m, Lake Vanda is hypoxic. Still farther down, the lack of any microbial mats in the hypersaline zone means that the small concentration of oxygen remaining in the bottom waters is not consumed, and the depths never become totally anoxic.

Among the Antarctic lakes, Lake Vanda has the smallest primary production, but it possesses the greatest biological diversity. Even so, the only primary consumers that have been found in this lake are a few species of ciliates. In summer, the edges of the five-meter-thick ice cover melt, producing a moat. The summer moat is dominated by a mat of filamentous blue-green algae, while under the ice, in the lake itself, there is a sequence of mats. At shallow depths, the dominant phytoplankton in this lake are two species of golden-brown algae in the genus *Ochromonas*. The golden-brown algae are the mince pie of algal taxonomy; they contain anything that cannot be linked to other algal groups. Some of these algae are ameboid, while others, like *Ochromonas*, are flagellated and are strong swimmers. Many are fresh-water algae, but the marine genera are an important component of the nanoplankton, the plankton that falls in the 0.5 to 0.05 micron size range. One of the two *Ochromonas* in Lake Vanda, *O. minuscula*, is small even among *Ochromonas*. Their small size helps these algae stay in the upper reaches of the lake, even without the aid of wind or waves.

Only a few diatoms are found at shallow depths in Lake Vanda. With increasing depth, the sloping lake bottom contains a succession of types of mats. Where the light is fairly bright, the blue-green algal mats (mainly composed of *Phormidium* and *Lyngbya*) take on a life of their own and behave like a living skin on the bottom. At these relatively well-lit depths the mats photosynthesize actively, and they get a pimply appearance because of trapped oxygen bubbles. The algae that make up these mats are filamentous and move in the characteristic gliding fashion of the blue-greens. They are also phototactic and glide toward the light in increasing numbers as the light intensity becomes stronger; this biological response enhances the pimpliness of the mats if the light intensity is high enough. Deeper than the pimply zone lie limp, prostrate mats that are still mainly composed of *Phormidium* and *Lyngbya.* At these depths these blue-green algae can barely make ends meet through photosynthesis. At a depth of 40 meters the dominant organisms are small *Phormidium*-like blue-green algae. Still farther down, the salinity starts to increase rapidly, and at about 55 m we encounter a salt zone with few signs of life, even though oxygen and light are still available.

The algae in Lake Vanda and similar lakes excrete part of the carbon they fix as water-soluble compounds such as *glycollate*; this action is tied to the high oxygen concentrations found dissolved in the water column. It is not clear what advantage (if any) the algae receive from excreting part of their fixed carbon, but it is easy to see the effect of these dissolved substances on other denizens of these lakes. For instance, Lake Vanda has a resident nonphotosynthetic dinoflagellate that probably absorbs these substances and utilizes them as sources of energy and carbon.

Another Wright Valley lake, Lake Bonney is composed of two lobes; the better-studied east lobe, with a bottom conductivity over three times that of seawater at the same temperature, is saltier than the west lobe. This lake is quite a bit shallower than Lake Vanda, but it is not as oligotrophic and almost no light reaches the bottom; however, because of its saltiness, the bottom is well oxygenated. Since almost no light reaches these depths, the numbers and types of organisms are few, and most are there not by design but by the merciless hand of gravity. In Lake Bonney, the blue-green algal mat ends abruptly at a depth of 14 meters. The next deeper zone in this lake is made up of a living coat of diatoms on the lake bottom. This zone is unique to Lake Bonney, and it shows how each of these lakes has developed a unique personality over the valley's 100,000-year existence. From a depth of 16 meters and down, the bottom of this lake is covered by a white flocculent precipitate that is most likely of mineral origin. At these increasingly saltier depths, only a few diatoms are found on the lake bottom.

In other Antarctic lakes located in other valleys, the deepest living layer is composed of flocculent mats of anaerobic blue-green algae. When light intensities become very low, these algae become anaerobic by uncoupling the two photosystems that make up oxygenic photosynthesis and reverting to the anoxygenic photosynthesis of their ancestors.

In lakes like Lake Fryllen, the abysmal fall to the anaerobic realm happens over a mere 10 m. At that depth the amount of Photosynthetically Active Radiation (PAR) is one hundredth of that on the surface. The only organisms that seem to have the right stuff to photosynthesize in this darkness are a few diatoms. These algae-in-a-pillbox lack an efficient locomotory apparatus, and they must depend on the manufacture of oil droplets to maintain their position in the water column.

The Oceanic Littoral Environment

In many respects the most benign antarctic environment is that found beneath several meters of seawater. Beneath the ice in winter the water temperature never drops below -2.8°C, and even though cold, this is a stable environment of little light or motion. In this realm there are two types of creatures. The top echelons of the food chain are occupied by warm-blooded intruders from above that harvest the bounty of this space with agility and speed and return to the surface, inexorably tied to the realm of light and air. The bottom echelons are occupied by the native invertebrate grazers and filter feeders such as sea urchins, sponges, and cnidarians. These organisms are in turn preyed upon by several species of sea anemones, sea stars, and ribbon worms. For these invertebrates, nutrients ultimately come mainly from the algae that grow on the undersurface of the ice, and also from the feces of the more active warm-blooded predators that rain into this realm from above.

9 Temperate Forests in Winter

Up to now we have visited places where a combination of cold and drought creates conditions under which only a few highly specialized organisms can survive. In temperate forests, on the other hand, there is enough moisture, and the temperatures are relatively mild during the growing season. These places are well vegetated and trees abound, but in winter the situation changes and what was thriving and green a month before loses all semblance of life. The curtain has come down and the actors have retired, but in a few places, new, smaller actors take the stage, now able to once again see the sun. I am speaking of those few winter-active plants that find freedom from competition for resources in winter, when the sun once again reaches the forest floor, and trees, now dormant, do not steal the soil's moisture as soon as rain drops hit the ground. As before, we will proceed from the northernmost regions where winter frosts are too severe and winter snow cover too extensive for any plant to be active, to more southerly climes where snow is short-lived and never totally covers the ground.

Winter Cold

Temperature affects different enzymatic reactions differently, and a temperature lower than the optimum, even though it may be above freezing, can still cause damage to a plant. A complicating factor is the timing of the cold spell. Plants are cold-sensitive to differing degrees during different parts of their life cycle. Seedlings are more sensitive than adult plants, and fruiting or flowering plants are more sensitive than dormant plants or plants already preparing for dormancy. If the local environment is such that frosts out of season rarely or never occur, plants can prepare well for even the lowest temperatures, and there is no cold on Earth that can harm such dormant plants.

Supercold winters are found inside continents; at equivalent latitudes, a larger continent will have colder winters in its heartland than a smaller one. For plants, another factor found at the heart of continents is the yearly fluctuation in temperature. In North America, a record low temperature of -63°C has been recorded in the village of Snag in the foothills of the Nutzotin Mountains in the Yukon (646-meter elevation). There are colder places, however. In the Cherski mountains of northeastern Siberia, near the village of Oimyakon, the winter minimum temperature has dipped to -68°C, whereas the summer maximum can reach 37°C. This is a yearly fluctuation of 105°C! (The amazing cold here is surpassed only in the Antarctic during its bleak winter.) Yet these areas of North America and Asia are well vegetated with conifers and other trees (even though the number of tree species is small). These forested areas are part of the taiga—the band of forests that extends around the globe just north of the deciduous woods and south of the tundra. Here a few species of conifers have developed mechanisms to protect themselves against extreme winter cold. Many other organisms, from the beetles that burrow in the trees' heartwood to the wood frogs that breed in the ponds, have evolved mechanisms to withstand extreme winter cold in these forests.

The northernmost taiga is a patchwork of thin trees dispersed among grassy meadows and in many ways reminiscent of the parklands often seen close to alpine timberlines. During most winters, temperatures drop below -40°C. Summer temperatures are not, however, very much different from those of richer forests farther south. The explanation for the northern taiga's depauperate flora is complex. At least part of the answer lies in the different frost-protective mechanisms found in different groups of plants. The many plants that can withstand prolonged exposure down to about -40°C, but not below, might be exploiting the supercooling of water to remain unfrozen. Relatively few plants are able to withstand harsher frosts. This dichotomy also seems to apply to animals.

Cold-Tolerance Mechanisms

Ice is the thermodynamically stable state for water at one atmosphere of pressure and less than 0°C. This statement says nothing about how fast this solid state is reached. In the laboratory, if water is extremely pure and one takes care to prevent sudden vibration, water can be cooled down to -39°C without undergoing freezing. (At this point, even tapping the beaker could create nucleation centers and set off almost instantaneous freezing.) This phenomenon is known as the supercooling of water. Supercooling is a kinetic effect, and not a thermodynamic lowering of the freezing point like the one produced by antifreeze compounds. Obviously the liquid inside living cells and tissues is not pure water, and supercooling organisms actively prevent the formation of nucleation centers, or mask them if they are present.

Those plants that can withstand temperatures lower than -40°C supercool to some extent, but in addition they are able to shut down cell function by a process similar to, but more sophisticated than, freeze-drying. This process (sometimes called equilibrium freezing) is not completely understood, but it involves removing water from the compartments that are most susceptible to ice crystal damage and letting these compartments dehydrate to the point where no free water (water that is able to turn into ice) is present. Of course, at this point these compartments cease their useful function, and the plant must shut down in a genetically defined and preprogrammed way in order to avoid injury. In order to shut down without injury, a plant must have time to complete this shutdown process, but while in this process, it is still susceptible to damage by cold. Plants get around the problem by using environmental cues that happen weeks or months before the onset of the deep freeze. To trigger the process, some plants (for example the northernmost opuntias) use as a cue the shortening day length, while others use the first frost of the season. Most trees near Snag stop growing when the day length drops to less than 20 hours, which in this area happens in late July, plenty of time to prepare for the first frosts. Table 1 shows several well-known trees, some deciduous and some evergreen, that can withstand the coldest temperatures both science and nature can muster.

Table 1 Some trees with cold-hardiness below -70°C.

SPECIES (COMMON NAME)	FAMILY (SUBFAMILY)	HABIT
Larix dahurica (Dahurian Larch)	Pinaceae (Abietoideae)	Deciduous
Larix laricina (Tamarack)	Pinaceae (Abietoideae)	Deciduous
Larix sibirica (Siberian Larch)	Pinaceae (Abietoideae)	Deciduous
Picea glauca (White Spruce)	Pinaceae (Abietoideae)	Evergreen
Picea mariana (Black Spruce)	Pinaceae (Abietoideae)	Evergreen
Picea obovata (Siberian Spruce)	Pinaceae (Abietoideae)	Evergreen
Pinus banksiana (Jack Pine)	Pinaceae (Pinoideae)	Evergreen
Pinus sylvestris (Scotch Pine)	Pinaceae (Pinoideae)	Evergreen
Thuja occidentalis (Eastern Arborvitae)	Cupressaceae (Cupressoideae)	Evergreen

Table 1 is only a partial list; another eighty pine species have been tested, and of these, nine have survived to -70°C. Since there are so many plants that can achieve this deep freeze, it is clear that the genetic mechanisms are widespread, and that some plants, given adequate warning, have no difficulty handling extreme cold.

For technical reasons, experiments in cold hardiness are seldom carried out below -78.5°C (the condensation temperature of carbon dioxide at one atmosphere), since lower temperatures require sophisticated, expensive equipment. This type of equipment is much more likely to be found in a medical cryogenics laboratory than in a forestry lab. The few studies that have been done at temperatures lower than -78.5°C have shown that cushion arctic-alpine species such as the purple saxifrage (*Saxifraga oppositifolia*), the moss campion (*Silene acaulis*), and diapensia (*Diapensia lapponica*) are absolutely frost tolerant during winter, withstanding liquid nitrogen temperatures of -196°C.

In the previous chapter we encountered the concept of the heat sum. It is sometimes easier (especially for trees) to obtain the heat sum for leaf expansion instead of the heat sum for seed production. The evergreen conifers of table 1 have, for leaf expansion, heat-sum requirements of around 950 degree-days, whereas the deciduous conifers have heat sum requirements of around 600 degree-days. Woody shrubs such as willows need just over 200 degree-days for leaf expansion. In the Canadian Arctic, the 650 degree-day line translates roughly into a line where the average July temperature is 10°C. This isotherm coincides (in most places) with the transition zone between the taiga and the tundra. Some trees can grow north of this boundary, but if a tree or shrub does not receive its full heat dose, it may not be able to "harden" before the onset of winter. In such a case, an individual of a species that would normally be winter hardy may succumb to low winter temperatures and desiccation. This may be an explanation for the lack of trees, other than the white spruce (*Picea glauca*), in the St. Elias range of the Yukon.

The Winter Survival Strategies of Insects

Unlike glaciers and permanent snow fields in high mountains, the boreal forests are only seasonally cold, but the cold season can be brutal. These forests are found in the middle of continents, and life in them takes on a pronounced seasonal quality. A common method used by insects to survive such winters is to enter a state of suspended growth called *diapause*. Insects enter diapause after being exposed to short day length and low temperatures, and in this state many insects can survive extreme cold.

Diapause is an immutable event; it is programmed to occur at a particular stage in the insect's development. It has been argued that the cold tolerance seen in diapausing insects is a separate phenomenon from diapause itself, and that it should be treated separately. I will follow this advice in these pages.

A survey of temperate arthropods shows that the least cold-hardy group, in terms of the minimum temperature that may be survived, is the freeze-intolerant group. This group includes mites (arachnids) and springtails (proto-insects), as well as many insects. If an insect is freeze intolerant, then it must use supercooling and behavior to avoid freezing. Behaviorally, flying insects can migrate as some butterflies do; other insects opt for much shorter trips and migrate vertically into the ground or under snow. In either case the temperature will be from 10 to 30°C higher at a half-meter depth than at the surface. In this way the pupa of the cabbage root fly (*Delia radicum*) overwinters in England without actually experiencing a single subfreezing night, whereas on the snow surface, freezing temperatures might occur 100 times in a year.

The mechanisms insects use to tolerate subfreezing body temperatures are varied. Sometimes different populations of the same species use different mechanisms, and sometimes even the same population uses different mechanisms over a period of several years. There is no universal agreement on the details of these mechanisms, but a broad picture can be sketched out. A simple classification of insects based on these mechanisms of protection to low temperature produces two groups.The first group uses the mechanism of supercooling. There are several adaptations used by insects to assure that supercooling is reliable enough to be depended on for winter survival. First, an insect must synthesize a substance that will produce the desired supercooling effect when dissolved in water without affecting critical biochemical reactions. Glycerol, a polyhydroxy alcohol (polyol for short), is such a chemical. It is a "biocompatible" solvent in the sense that it does not affect enzymatic reactions, as high concentrations of other sugar alcohols can. It is also an intrinsic part of the biochemistry of most cells; therefore, almost every living thing on this planet can produce it and regulate it. Glycerol is composed of a three-carbon backbone and three hydroxyl groups, one attached to each carbon. The beauty of glycerol is its highly polar nature and hydrogen-bonding ability. Its many other advantages include its pronounced supercooling properties.

An antifreeze is a substance that, when added to water, will lower the temperature of equilibrium crystallization. Automobile antifreezes are based on this property. If a glycerol and water solution is maintained at any temperature above its equilibrium freezing temperature, it will not freeze even if it is seeded with ice crystals. If the temperature is dropped further, below the freezing temperature of the solution, the solution will not freeze unless some nucleation sites are added or are caused to form. This metastable state is the supercooled state, and glycerol has the uncanny ability of lowering the limit of the supercooling of water twice as effectively as it lowers the freezing point. Since life itself is a neatly choreographed dance of metastable states, supercooling fits right in. Supercooling is a crapshoot; it might work for only a day, and then, by some random motion, a set of water molecules find themselves forming a neatly arranged structure called ice. However, this might not happen for a month or for six months. Insects have figured out how to load the dice.

Supercooling insects have proteins that can arrest the development of any nascent ice crystal. The surface of a growing ice crystal recruits water molecules from the liquid phase in contact with the ice; the proteins attach to the ice surface and block the recruitment. As long as there are enough proteins in solution, all nascent ice crystals will have arrested growth. If the temperature drops further, however, more spontaneous nucleation centers will form, until some will escape inhibition by the proteins and freezing will progress uncontrolled. These proteins should properly be called freezing-hysteresis proteins, but the term antifreeze proteins has come into popular use, even though the mechanism is completely different from freezing-point depression. There are several extra precautions taken by some of the insects that use supercooling as their protection mechanism. One of the most common precautions found is defecation. By voiding its gut, the insect removes foreign material that may be the source of nucleating agents that will compromise its supercooling. Any moisture on the insect's surface may cause ice crystals to spread like an infection through the weak permeable areas of the integument (surface of the body). Some insects preparing for overwintering seek out dry sites; some others also decrease their body water content by twenty to forty percent. These precautions help the insects avoid any source of nucleating agents.

Some insects, such as wood-boring beetles and gall-forming insects, must spend the winter in the plant organs they infest. In such an exposed condition, temperatures can drop to nearly the air temperature. In the boreal forests of Alaska, the Yukon, and eastern Siberia, this means surviving minimum temperatures of -60°C where supercooling alone won't pass muster.

An insect must tolerate freezing when the environment is so cold that freezing will happen no matter what protective measures the insect takes. Freeze tolerance seems to be the most often used adaptation of pterygote (winged) insects in extremely cold places. A freeze-tolerant insect actually stimulates freezing by producing nucleating agents that will cause water to freeze at close to 0°C. The trick here is the same one we found in extremely cold-hardy plants. Freezing has to be induced outside the cells; in insects this means the hemolymph. By producing nucleation proteins in the hemolymph, insects can control the nucleation of ice and its growth. If ice growth can be limited to the extracellular spaces of the hemolymph, it is thought that the organism will be spared any injury. Again, as in plants, ice formation outside cells causes a gradient that allows water movement into the hemolymph, and as the cells dehydrate, their increased solute concentration prevents them from freezing. However, ice might form in the insect's gut before it forms in the hemolymph, if the insect has ingested nucleating agents in its food. And ice formation in the gut of an insect can prove fatal. In such a scenario ice may penetrate the gut wall and cause havoc, as rampant crystal formation can then take place. As in the case of supercooling, some insects seem to take care of this possible threat by voiding their guts, while others do not take this precautionary measure and yet survive. Simply controlling the place of ice formation is not enough, however. As water leaves the cells, their solute concentrations increase. Some of these solutes are toxic to the cells themselves. What the insect needs is some means of diluting these dangerous compounds back to the concentrations found before ice formation began. Whatever is used for dilution must be nontoxic; it must remain in solution at low temperatures; it must be easily made and destroyed by the cells' biochemical processes, and it must easily penetrate internal cell membranes. If it were not to penetrate internal cell membranes, then every cellular compartment would have to make its own supply to protect itself from increasing toxic levels of other solutes. There are only a few compounds that fulfill all of these requirements—one of the most common is our old friend glycerol. Therefore, glycerol is used by both supercooling and freeze-tolerant insects.

A Tour of Winter Life

We will start our tour of winter life in the temperate deciduous forests of North America, where many unassuming early spring bloomers often turn a solitary walk in the woods into a memorable experience. From here we will go to the Mediterranean climates, where water, more than cold, dictates the season of activity.

Northcentral North America

While the rulers of the forest sleep, some of the lesser denizens of the forest floor seize their chance to capture the light that is stolen at other times of the year. The opportunity arises in areas with winters severe enough to stimulate most woody vegetation to go dormant, but with winter snow cover infrequent enough to allow herbs and low shrubs to bask in bright winter sunshine. For example, in southern Illinois and southern Indiana the ground is covered with snow for only ten to thirty days in winter; therefore, more than half of winter is snow free. (Snowy conditions increase toward the north in this area.)

In eastern Indiana the procession starts in early and mid February with the short-fruited whitlow-grass (*Draba brachycarpa*), the toothwort (*Dentaria lasiniata*), and the skunk cabbage (*Symplocarpus foetidus*). These plants range widely through eastern North America and into the Midwest, but the skunk cabbage also exists as disjunct populations in Siberia and Japan. The toothwort gets its name from the toothlike shape of the small underground tubers that form a ring around its root system. It grows in moist shady woodlands where it forms large patches of small white flowers washed with pink. Like in the short-fruited whitlow-grass and many other members of the mustard family, the tubers have a peppery taste. The earliest-blooming of the three plants is the skunk cabbage. It can forge its way up to the surface using its heat-generating ability; the other two, which are devoid of this rare skill, must wait for the ground to be free of snow for several days before starting the flowering process.

In southern Illinois, all of the plants that flower in late winter do so in deciduous woodlands where there is a definite advantage to an early start to

The skunk cabbage (*Symplocarpus foetidus*).

flowering. These plants have either evergreen leaves that are held through the winter, or leaves that unfurl after the plants have flowered. Still, there are several problems with flowering so early. For one, the chance of frost damage to the flowers is great. Some of the plants deal with this threat by producing flowers that are amazingly resistant to frost damage, whereas others simply take their losses and quickly produce new flowers. In their protected locality in rich woods, these plants probably do not experience the same severity of weather that afflicts people a couple of meters above the ground, where night temperatures of -20°C probably happen several times each winter. The temperatures experienced by some exposed plants are not too much milder than that, yet some of these plants seem to shrug off the previous night's frost and cheerfully thaw with the morning sun. In more northerly climates it becomes harder and harder to flower and start growth in late winter, not only because the frosts are more severe, but also because the blanket of snow smothers all aspirations. It may be that the difficulty imposed by snow is much more severe than that imposed by colder temperatures, but separating the significance of the two factors is best done in the laboratory. Since these plants are in a race for first place, it is not surprising to find that many of them have bulbs, corms, or tubers that store the sugars necessary for this early spurt of growth.

Another late-winter bloomer, spring beauty (*Claytonia virginica*), puts on a splendid show in late winter and very early spring. Sometimes colonies of this plant form masses of light pink on humus-rich woodlands. In the fall, the spring beauties put forth their new leaves, which photosynthesize away until the end of spring. In late winter and spring the pink flowers appear on short stems; as the stems grow, the flowers follow the sun, closing at night and in periods of cloudy weather. By the time the plant sets seed, the flower stems are over 30 cm tall. In June the spring beauties vanish from the scene, leaving no trace that they ever graced the woodland's floor. Alas, overactive gardening enthusiasts have diminished the wild population of spring beauties to such an extent that such a sight is very rarely encountered any more. Luckily for both the plant and those of us that can appreciate its beauty better in the magic solitude of the winter woods, there are several beautiful cultivars available in the nursery trade, and I hope that domestication will allow this plant to again flourish in the North American woodland.

The false rue-anemone (*Isopyrum biternatum*) belongs to the buttercup family, whose members are so common in mountains and polar lands. It is a dainty plant difficult to distinguish from the true rue-anemone and anemones in general. Its most distinguishing characteristic is the pearl-like flower buds that may be the earliest sign that spring is approaching. A relative, the liverleaf (*Hepatica acutiloba*), is so called because the leaves produced in the spring progress from hairy and light green to olive green and finally, after persisting through the winter, at the end of the season they turn purple and more or less lobed like a liver.

The Skunk Cabbage

Cold is a relative thing. If you are from the tropics, winters in peninsular Michigan are brutally cold and unbelievably snowy. So it is for plants. However, a family of tropical plants has made it to northern Michigan; these are the aroids. Some, like *Dieffenbachia* and philodendrons, have made it there as part of the sometimes pampered indoor flora. Some others have made it on their own. One of these, the skunk cabbage (*Symplocarpus foetidus*), has adapted a family trait to an unusual purpose, and in so doing has taken the trait to extremes. The inflorescence in aroids is a complex structure composed of a leaf-like spathe and a spadix; in tropical forms, the spathe varies from white to bright red, but in temperate forms it is a drab green or brown, or even absent altogether. The true flowers are found on the spadix, which can be small and simple as in the wild calla (*Calla palustris*), or gigantic and complex as in *Amorphothalus titanum* (an accurate mental picture can be derived from the Latin name). Many aroids can raise the temperature of the spadix by a few degrees. In the tropical rain forest this trait helps these plants get more of their chemical messages into the still air and improves their chances of being pollinated. In the crowded confines of the rain forest where competition is fierce, every little edge helps. A classic experiment carried out by Fritz Knoll in 1926 demonstrates this vividly. He created a beautiful purple glass replica of the inflorescence of the voodoo lily (*Sauromatum guttatum*). He then inserted small light bulbs as a source of heat in some of the glass flowers. Some of the heated and some of the unheated glass flowers were also painted with putrefied blood to simulate the wretched odor of the real thing. Simply warming uncoated flowers attracted no insects, but the coated flowers attracted some, and the coated and heated flowers attracted the most.

In the deciduous forests of eastern North America, the skunk cabbage has become a master of heat production. In late winter, while the leaves are still off the trees, but the air temperatures have started their gradual warming trend, the skunk cabbage turns on the heat, not a few degrees (as other aroids), but an astounding 20 to 35°C. It keeps this up for almost a month in late winter, maintaining a nearly constant flower temperature of 23°C, as comfortable as your living room, and just as flies are attracted to the inside of heated houses in spring, insects of all kinds are attracted to the skunk cabbage's little parlor. Heat production for this plant has other distinct advantages. It melts the covering of snow, making the plant easily accessible to would-be pollinators and able to receive the warming rays of the sun through the leafless tree branches (rays which raise the temperature of the flower another 4°C). It also speeds up development of flowers and seeds.

Heat production is also found in five other families, ranging from palms to cycads, but in these tropical plants heat generation is only used to help volatilize pollinator attractants. At least in the aroids, it has been shown that the process of heat production is under the control of a newly discovered plant hormone—salicylic acid. Salicylic acid is the chemical precursor to aspirin, which has several physiological effects on mammals. In plants, salicylic acid is important in heat production and in protection from infection. Injecting aspirin into an aroid flower will in fact cause its temperature to increase. Using heat to beat out the competition and mature seeds faster is one of the overriding themes in cold environments, but the skunk cabbage is one of the few plants that generate their own.

Fragrance and Color in Early Spring

The fragrant sumac (*Rhus aromatica*) exemplifies one of the most curious qualities of early blooming shrubs, their strong fragrances. This trait can be understood as an adaptation to an environment where pollinators are few and far between, and plants must use long-range advertisements in order to gain the attention of the few misguided insects that venture from their protected winter homes. Another characteristic worth noting in these flowers is the lack of color—white or light pink are the only colors that abound in these woods at this time. The white or light-colored flowers are geared to the few pollinators available, mainly flies. Colors such as gaudy reds and blues will not be displayed until the beetles and other insects that respond to these colors awake. Other shrubs that have taken this evolutionary road include some of the witch hazels and the heaths.

Animals in Northern Temperate Forests

Many mammals and birds are active through the winter in temperate woods, but only a few cold-blooded vertebrates have learned how to survive very cold winters. Cold-blooded vertebrates have the disadvantage that their mass has to be heated by the sun. In North America the only cold-blooded vertebrate to reach the Arctic Circle is the wood frog (*Rana sylvatica*). It reaches its northernmost extent along the Mackenzie River in the Yukon, and it may actually reach the arctic shore along the Mackenzie River delta. It may also reach the tundra proper along the Ungava peninsula of eastern Canada. This frog survives extreme winter cold by allowing 35%

of its body water to freeze and protecting the rest with high concentrations of glucose.

High concentrations of glucose may protect tissues from direct freezing damage, but there are also serious but indirect effects of freezing. In order for the tissues to remain undamaged as the heart stops beating and the blood congeals in the arteries these frogs must have a well developed anaerobic metabolism. They also must be able to produce high levels of reactive-oxygen-destroying enzymes to protect themselves as they emerge from freezing and anoxia.

The wood frog has also developed behavioral adaptations to cope with the cold. In early spring (May in the far north), after heavy rain or maybe after a strong thaw (no one knows for sure), the frogs emerge from their wintering places and migrate to their breeding pond. Most frogs in a region arrive at the breeding ponds at the same time. Egg laying is not random, and communal egg masses are common. These masses might contain the eggs of up to one hundred females. Since each female may lay one to two thousand eggs in a tennis-ball-shaped mass, the overall communal mass can be considerable. This behavior might provide thermal advantages to the eggs, since the center of such masses can be up to 1.6°C warmer than elsewhere in the pond.

The eggs of the wood frog have several structural adaptations that increase their heat-gathering efficiency. Eggs of amphibians have a protective gelatinous covering called the mucoid capsule; this covering can fit snugly over the zygote or it may be large in relation to the zygote. When the egg is first laid, the capsule swells depending on the ionic concentration of the pond water. The capsule serves several protective roles. It prevents fungal penetration, and it deters some would-be predators. But probably its most important role in cold climates is that it increases heat retention by the egg. With larger capsules, heat retention is much greater. This larger transparent mass gives the egg extra insulation while still allowing it to collect solar energy. Frogs that lay their eggs in sunny locations have pigmented eggs; the wood frog in particular has a darkly pigmented side that always points up. This side aids in the heating of the egg mass, but in the case of the wood frog there is also an unexpected assistant—an alga grows within the confines of the capsule, and its dark color helps to increase the temperature of the egg. The full relationship between the alga and the frog has not been explored.

When winter comes, wood frogs find cavities in which to hide; these might be under rocks, leaves, logs, or in shallow burrows. In some of the wood frog's range, freezing is a real possibility, and this frog does produce substantial amounts of glucose and can stand freezing to at least -6°C. Other amphibians—such as the Siberian newt *Salamandrella keyserlingii*—also accumulate glycerol, and in so doing, possibly avoid the deleterious effects of very high glucose concentrations. Some ecotypes of these freeze-tolerant amphibians probably can survive prolonged periods at -6 to -10°C.

Many of the animals we dealt with as winter residents of the Arctic are also winter residents of the northern temperate forests. Other winter residents are migrants from more northerly lands, while still others are forest residents year-round. For winter-active animals (those that do not overwinter in a torpid state) the difficulty is one of lack of food. The low temperatures bring to a halt arthropod activity, except in the warmer areas where a few flies such as crane flies can still be found on warm days. These flies (and in some places, night-flying moths) also seem to be the only pollinators available to winter-blooming plants. Along with the flies, active fly predators such as spiders are present in these woods, but their adaptations remain unexplored.

Southeastern North America

Besides the forest perennials we have talked about, there are several annuals that have the ability to withstand frosts during growth and flowering, and in places with mild winters, these plants may flower all year long. These plants are normally considered weeds of open places, but in the middle of winter, any flower is a good flower. Notable examples of this type of plant are the annual bluegrass (*Poa annua*), the common chickweed (*Stellaria media*), the black medic (*Medicago lupulina*), and the red dead nettle (*Lamium purpureum*). The red dead nettle is one of the earliest herbs to bloom in the grain fields of Europe, and because it can survive a mild winter it may produce three generations a year. The closely related henbit (*Lamium amplexicaule*) is also a cosmopolitan weed which can be seen blooming in the middle of winter from southern Indiana to Florida. I have observed it in bloom even when night temperatures drop to -15°C. Lower temperatures damage the flowering stems, but seem to leave the young nonflowering stems unaffected. The same thing can be observed in the black medic and the annual bluegrass, but these seem to be killed outright by temperatures below -15°C.

The genus *Lamium* is actively evolving, and in Washington state a new *Lamium* was recently born. Exactly when this act of evolutionary slight of hand occurred is not known with certainty, but the red dead nettle and the henbit, in a bit of unbridled promiscuity, have produced a handsome and vigorous offspring that is bound to outdo the exploits of its parents. Most such interspecies offspring are sterile because of incompatibilities in the number of chromosomes and also in the timing each species uses to step through the motions that lead to monoploid sex cells. Sometimes, however, such normally sterile progeny undergo a rare cellular accident. This accident goes by the name of allotetraploidy and is caused by the incompatibility of the two sets of chromosomes. Normally the number of chromosomes in sex cells is halved (a process known as meiosis), so that the union of a male and a female gamete will then reconstitute the normal diploid complement of chromosomes. In some of the progeny produced by this cross, this halving did not occur, and the sex cells produced by these plants contained the full diploid complement of chromosomes. When these cells united, the result was a plant with four complete sets of chromosomes, a tetraploid. This dead nettle has been christened *Lamium hybridum* and is much more aggressive than either of its parents; it will soon be coming to a pasture near you.

Mediterranean Climates

The selective pressures that drive forest floor herbs in temperate climates to take on a late-winter or early-spring blooming habit are more of an opportunity than a necessity, and since nature does abhor a vacuum, a few such plants are bound to exist in many deciduous woodlands around the world. In some places on the globe, plants bloom in winter, not because they are seizing a little-used season for themselves, but because the alternative is too harsh to even attempt. These places have Mediterranean climates, in which the winters are wet and cool, but the summers are hot and in some places waterless. The typical Mediterranean climates include parts of the western coasts of North and South America, parts of southwestern Africa and Australia, and of course, the Mediterranean basin.

Some areas maintain the typical seasonality of precipitation of Mediterranean climates but have more pronounced fluctuations in temperature. These places include Asia Minor and parts of the Middle East. These lands are parched during the summer when little rain falls. An example is Jerusalem, where rain is so rare in July that as of 1980, after more than one hundred years of record-keeping, the first July shower had yet to be recorded. In such a land, shrubs green in winter and lose their leaves at the coming of summer. In lands with a similar but somewhat colder climate, winter growth is slowed and all the activity is concentrated in the spring.

The characteristic growth form of interest to us here is the geophyte—a plant that survives the dry period by leaving the parched surface and lying

Table 2 Some Winter-Flowering Perennials

Scientific Name	Family	Common Name	In Bloom	Hardiness
Chionodoxa	lily			
luciliae		glory-of the-snow	Mar.	-40°C
sardensis		glory-of-the-snow	Mar.	-40
Scilla	lily			
biflora		squill	Mar.	-35
siberica		Siberian squill	Mar.	-35 to -46
tubergeniana		squill	Feb.	-29 to -35
Crocus	iris			
ancyrensis		crocus	Feb.	-40
chrisanthus		crocus	Feb.	-35 to -40
sieberi		crocus	Feb.	-23 to -40
susianus		crocus	Feb.	-35 to -40
tomasinianus		crocus	Feb.	-29 to -40
Iris	iris			
danfordiae		reticulated iris	Mar.	-23 to -29
histrioides		reticulated iris	Mar.	-23 to -29
reticulata		reticulated iris	Mar.	-23 to -29
Galanthus	amaryllis			
caucasicus		snowdrop	Dec.	-23 to -40
elwesii		snowdrop	Jan.	-35 to -40
nivalis		snowdrop	Feb.	-40
ssp. *cilicicus*		snowdrop	Jan.	-40
Leucojum	amaryllis			
vernum		spring snowflake	Feb.-Mar.	-35
Narcissus	amaryllis			
asturiensis		daffodil	Feb.	-35
Eranthis	buttercup			
hyemalis		winter aconite	Jan.	-23 to -35
Helleborus	buttercup			
foetidus		stinking hellebore	Dec.	-23
niger		Christmas rose	Jan.	-29
orientalis		lenten rose	Feb.	-29
Iberis	mustard			
sempervirens		snowflake	Feb.	-35
Cyclamen	primrose			
cilicium		cyclamen	Oct.-Nov.	-18 to -23
coum		cyclamen	Jan.-Mar.	-18

dormant at some depth where its stores of food and water are relatively safe. Lands with Mediterranean climates and other, even drier, lands are the cradles of many of the commercially available winter-flowering plants (see table 2 for a list). Among the most celebrated of these are members of the amaryllis family (Amaryllidaceae) such as the snowdrops (*Galanthus* spp.), especially *Galanthus nivalis*. Snowdrops evolved in the northern Mediterranean region in relatively moist woodlands such as floodplains, but they can also grow at modest elevations.

The other undisputed aristocrats of the winter landscape are the members of the buttercup family such as the various hellebores, especially the Christmas rose (*Helleborus niger*) and the Lenten rose (*Helleborus orientalis*). The center of evolution for the hellebores is the Mediterranean basin, Anatolia, and the Caucasus. There are altogether about 20 species, including some endemic to the islands of the Mediterranean Sea such as Corsica and the Balearic Islands. In Europe, they are plants of limestone-derived soils of the foothills of the Alps, the foothills of the Apennines, and those of the southern Carpathian mountains. The Christmas rose, like the snowdrops, is a true winter plant sending up flowering stalks from mid-December to March when the season is not too harsh.

Both the snowdrops and the hellebores react to very cold temperatures in the same way. During a hard freeze, when the night temperature drops to -15°C or below, the flowering stems droop to the ground as if mortally wounded by a myriad icy knives, but as the temperature climbs the next morning, the plants regain their upright posture and are once again the undisputed rulers of their frozen patch of forest.

Other less well-known members of the flora of Mediterranean climates are the various scillas (*Scilla* spp.) and the glory-of-the-snow (*Chionodoxa* spp.), all in the lily family. These plants bloom somewhat later than the hellebores or the snowdrops. A few of the scillas have managed to venture much farther afield than either of those two. They boast members like the Siberian squill (*S. siberica*), that originally grew in central Asia and Siberia, and the southern African scillas such as *Scilla pauciflora* and *Scilla socialis*, that look like miniature aloes. The scillas have undergone rapid evolution in the recent past, as is evident from a few endemics that inhabit tiny islands in the Mediterranean, such as *Scilla dimartinoi* on Lampedusa Island, an Italian island off the coast of Tunisia. The scillas and the glory-of-the-snow must have diverged relatively recently, since fertile hybrids between the two genera can still be produced. A hybrid is available commercially and has been christened *Chionoscilla*.

10 The Mountain Environments

Unlike the polar regions, where the clear enemy of life is unrelenting cold, each mountain range has its own unique set of villains. Among the most lethal of these are warm, dry air in winter and freezing temperatures during the growing season. On Earth, dry continental mountain ranges such as the Pamirs of Tajikistan have the widest daily temperature fluctuations. In the spring, nightly temperatures may drop to -20°C in these mountains, yet some wormwoods and cinquefoils are able to unfurl their spring leaves under these conditions.

Except for the most remote and rugged ranges, humans have had a tremendous impact on the plants and animals of the high plateau areas of the world. The wild animals that have historically called these places home are now found only in the most forbidding and lonely places, such as the northern Chang Tang, where the last herds of chiru and wild yak roam.

We are mainly interested in those mountains that pierce the upper limit of the biosphere, where environmental factors become so severe that few organisms can survive. In one sense, all mountain ranges with a significant area above the orographic snow line pierce this limit, since only algae and a handful of invertebrates can grow on permanent snow. Of more interest are those few mountain ranges where dryness makes the snow line retreat upslope, and there is a definite no man's land between the highest continuous vegetation and the snow line. On some mountains this area encompasses two distinct regions: one has been called the alto-alpine and the other the aeolian zone. These are the two zones of main interest to us here, but we will deal with the more benign forest and alpine zones when discussing particular aspects of life on mountains.

In this chapter we will first discuss the importance of daily temperature variations in the lives of high-altitude plants and animals. Then we will move from the northern polar region toward the equator, visiting several mountain ranges that have piqued the interest of researchers. Each place will clearly show some limit to life at work. A look at a topographic map of Earth shows many other mountain ranges that might yield important clues about the limits to life, but most of these have not been studied because of their remoteness.

Diurnal Temperature Variation

The High Arctic in the summer is definitely cold, but the average diurnal air temperature variation is small, amounting to just a few degrees. At Alert, located at 83°N latitude on Ellesmere Island, the difference between average "day" and "night" temperatures is a mere 1°C in the summer. This small temperature fluctuation is due to the ever-present sunlight, and it allows plants to enter a state of constant growth. This is not to say that the temperatures always stay within this range, since in mountainous areas the sun will be blocked and more pronounced fluctuations will exist. But on the whole, in the summer the temperature fluctuations in the High Arctic mountain ranges are benign compared to those in the Low Arctic and subarctic mountain ranges.

On the other hand, in high-latitude mountain ranges outside the High Arctic, such as the Romanzof and Franklin mountains in Alaska and the Icefield Ranges and St. Elias Mountains in the Yukon, temperatures can plummet well below freezing on every day of the year. In such circumstances, plants must maintain significant freezing tolerance but still keep metabolically active. This proverbial cake can be had and eaten too, but only by a few species. That some species can accomplish this feat shows that it is possible; that it is not seen more often may mean that the need only arises in a few localities on Earth. There have been a few studies of this ability in cushion plants such as the purple saxifrage and the moss campion. In the purple saxifrage, it seems that once a population breaks dormancy, most individuals go all-out and throw caution to the wind. These individuals are poised to complete their reproductive duties faster and produce more seeds, while a minority of individuals are genetically programmed to be more cautious and retain most of their rhizomes' cold tolerance throughout the growing season. This extra baggage slows the cautious individuals down and presumably diverts some of the resources they need to produce seeds, but it keeps them alive even if the night temperatures at the height of the growing season were to fall to -25°C (which is highly unlikely under current climatic conditions, but may not have always been so). Such freezes would kill unprotected plants outright, while protected plants would only lose their tender leaves and new shoots. In a variable environment, where in some years only a few freezes occur during the growing period, but in others that period is peppered with freezes, the population will stay mixed. Were the occurrence of severe early season freezes to increase, the plants that reproduce more slowly would start to dominate.

In the alpine areas of the Pamirs in Tajikistan, daily lows fluctuate from a few degrees above freezing in the summer to -50°C in the winter. Here, springtime night temperatures can drop to -20°C and below, but daytime temperatures of 30 to 40°C are not uncommon: a daily fluctuation of 50 to 60°C. The early-season daytime warmth is a resource not unlike water or light. The plant that can tap this resource gains a competitive edge in the community. Several rosette plants, among them a cinquefoil, *Potentilla multifida*, and a wormwood, *Artemisia skorniakowii*, have tapped this resource by starting growth in early spring while retaining a fair amount of frost tolerance.

Frost-Tolerance Mechanisms

To protect their growing tissues from hard spring frosts, some of these plants adopt supercooling mechanisms, while others are tolerant of equilibrium freezing. As we will see, some plants of tropical mountains opt for a supercooling approach, but once the temperature falls lower than their supercooling limit, they are damaged. As you recall from the previous chapter, in equilibrium freezing, plants allow ice to form outside their cells. In the compartments where the ice forms the equilibrium vapor pressure of water falls, producing a gradient of vapor pressure that allows water to move out of the still-liquid cells into the ice realm outside the cells. If this process is slow enough the interior of the cells remains liquid, while the ice continues to grow outside the cells. There are many questions still unanswered about frost-resistance mechanisms. For instance, pure water can supercool to -39°C, but at this temperature it will rapidly freeze. Is this limit also imposed on plants and animals? The answer seems to be that plants are constrained by this limit, but animals are not. What if springtime frosts were more severe than the limits of supercooling? It has been proposed that under such circumstances, mountain plants would minimize supercooling and simply go into equilibrium freezing. Theoretically, this mechanism should provide greater protection than supercooling when temperatures drop below a certain point; in practice this point seems to be around -11°C.

Equilibrium freezing is no panacea, however. There is no such state as "frozen solid"; there are critical amounts of water that cellular compartments must have—packing material for the fine china of cellular machinery if you will. This packing water is tightly bound to cellular components, but as the temperature drops, this water is more and more likely to be bound to the extracellular ice, leaving the cellular components to knock about and damage each other. For this reason, even being able to withstand freezing is no guarantee that an organism will survive as the temperature keeps dropping. In plants from tropical mountains the lethal temperature limit seems to be around -20°C. Another complication when using equilibrium freezing as a protection mechanism is the rate of cooling. At least for some animals, the cooling rate can be so fast that equilibrium freezing cannot be maintained; ice forms intracellularly, and the animal dies. For example, when several dozen larvae of the freezing-tolerant goldenrod gall fly (*Eurosta solidaginis*) were cooled to -40°C at 10°/minute, only 7% survived to adulthood. For plants, the question of how fast is too fast remains unanswered.

The rate of cooling can affect the survival of organisms in two other ways. First, any biochemical preparations organisms must make to survive the following night's frost take time; thus a slow cooling rate can be beneficial. For example, it has been shown that for the flesh fly *Sarcophaga crassipalpis*, to survive -23°C it must prepare by producing 2 to 3 times its normal level of glycerol. Preparation is induced by a drop in temperature to between 0 and 6°C, and it is not completed for another couple of hours after that drop. In general, if the rate of cooling is much greater than 3°C per hour, even supercooling insects freeze to death. The animals that supercool or undergo reversible freezing are for the most part small (there are a few exceptions such as the wood frog); some of the plants can be quite large, however.

The other effect of the rate of cooling on the survival of organisms is through the organism's thermal time constant, which is a measure of how well an organism's internal temperature tracks the environmental temperature. The thermal time constant of small organisms is short, and their core temperature tracks the environment's temperature with a lag of, at most, a few minutes. A long thermal time constant leads to poor environmental tracking—a good thing when you are trying to avoid exposing sensitive tissues to environmental extremes. It can be achieved by several separate but complimentary mechanisms: large size, low surface area, and good insulation. For a long thermal time constant to provide an adaptive advantage when the temperature changes every 24 hours, the organism must be larger than several kilograms and well insulated. Some mountain plants meet this requirement, and the temperature of their tender parts may never approach that of the environment. This thermal buffering is normally accomplished by storing water in different plant organs. Since water has a much higher heat capacity than air, an enclosed volume of water will drop in temperature much more slowly than an enclosed volume of air exposed to the same outside temperature. For instance, the inflorescence of *Lobelia telekii*, one of the giant lobelias that grow on the slopes of Mt. Kilimanjaro, is hollow and filled with four liters of liquid. Even when the air temperature drops to -12°C right before sunrise, the inflorescence never drops below 0°C. However, there is one difficulty with this adaptation: it is completely dependent on the size of the plant, and seedlings have no thermal buffer to speak of. This size advantage means that adult plants can survive when the climate turns deadly for seedlings, and therefore, relic adult populations can be found in several mountain ranges where there is no trace of juveniles.

Arctic and Temperate Mountains

Temperature-wise, the most extreme climates on Earth are the ice-cap climates. These climates are found in Antarctica, in Greenland, and on a couple of the largest ice fields in the northern continents, like Alaska's Juneau ice field. A few mountains are tall enough that their summits peak above the surface of the ice in each of these areas. Other mountains stand their ground while the edges of the ice sheets surge and wane around them, like a child in the surf. Still others are active volcanoes like Mt. Erebus in Antarctica, and keep the ice away with their own internal heat. All these mountains provide the first requirement for plant colonization: bare ground.

The importance of some of these mountains (and the ice-free surface they provided) as refugia during past ice ages is undeniable, but it is almost impossible to prove that any one isolated peak was the refugium that allowed a certain plant or animal to survive an ice age. These mountains (or nunataks, as the Inuit call them) provide differing grades of refuge to the organisms that inhabit them, depending on the severity of the surrounding climate. If the nunatak is on the coast, so that the ice sheet slides around it into the sea, a part of the coast will most likely remain free of ice, allowing both animals and plants to flourish in the relatively maritime climate. Not only is this type of refugium milder than the continental refugia such as Beringia (covered in chapter 8), but open water allows marine or amphibious creatures to flourish, tremendously increasing the biological diversity of these Noah's arcs in a sea of ice. Other nunataks are much less hospitable; they are completely surrounded by ice and take on the overall climate of the ice. These are the environments that concern us here.

On the east coast of Ellesmere Island there is a biological oasis called Alexandra Fiord; it has been extensively studied by scientists for the past twenty or more years. The lushness of this lowland extends only a short distance upslope into the mountains that surround the valley. In these uplands we will start our tour of extreme mountain environments. The mountains around Alexandra Fiord are not high; nevertheless, they are highly glaciated. In a few places here and there nunataks poke through the seams formed by converging glacier tongues. A few of the more accessible nunataks have been studied. Up to 1000 meters they show relatively sparse vegetation of a polar-desert character, including purple saxifrage (*Saxifraga oppositifolia*) and the woodrushes *Luzula arctica* and *Luzula confusa*, along with smaller amounts of grasses (mainly arctic bluegrass *Poa arctica*) and cinquefoils (mainly arctic cinquefoil *Potentilla hyparctica*). Above 1000 meters up to the crests of the mountain ridges (the crests are at about 1500 meters around Alexandra Fiord), plants are few and far between; the only conspicuous ones are the purple saxifrage and the arctic poppy, together with mosses (*Polytrichum* and others) and crustose lichens. Even at the highest elevation that has been surveyed (1400 meters), a springtail, *Vertagopus arcticus*, can be found feeding on windblown debris and whatever crustose lichens manage to survive at this elevation. This aeolian zone, where no plants live, but where small arthropods like the springtail seem even to thrive, is made possible by the agency of upslope winds that carry small bits of plant material or dead insects to the higher elevations. The upslope winds also carry seeds and other plant propagules. The reasons why these would-be colonizers do not manage to gain a foothold on these barren higher slopes are crucial to understanding the limits to life on mountains. To understand these reasons we must first have data about the environment and the would-be colonizers.

There is little climatological data for nunataks per se, but a few weather stations have been maintained on the glacier ice surrounding such isolated peaks. There were several manned weather stations in southwestern Yukon from 1963 to 1971. These stations were located in the Icefield Ranges and the St. Elias Mountains, which boast some of the highest peaks in North America, such as Mt. Logan (5951 meters), the highest peak in Canada. Although this mountain is not a nunatak (its flat summit is covered with snow), it gives an idea of the type of climate experienced on continental (as opposed to coastal) nunataks. The one weather station operated on Mt. Logan (for three summers between 1968 and 1970) was located on the western end of the large glaciated summit plateau at 5365 meters, making it the highest weather station ever operated in North America. In July, the average temperature was -18.3°C, the average low temperature was -23.9°C with an extreme of -34.4°C, and the average high temperature was -12.2°C with an extreme of -3.3°C. It was too difficult to recover weather records from this station during winter, and there is no data as to the extremes experienced then, but the lowest temperature ever recorded in North America (-63°C) was recorded 200 km northeast of Mt. Logan, suggesting that winter temperature minima are very severe on this mountain.

Some of the nunataks on the Icefield Ranges have been explored by biologists; vegetation on these mountains is hard to find, but it is there. Snowflush communities on the nunataks have a composition different from that of the polar snowflush communities discussed in chapter 8. That chapter states not only that snowflushes are cold and dark, but also that vegetation under heavy snows can be encased in ice and in the darkness can be exposed to prolonged anoxic periods, and therefore, not all deleterious effects can be attributed to the low temperatures or low light levels. Two plants that can be found at high elevations on the nunataks of the Icefield Ranges, hiding besides boulders or in snowflushes, are the umbrella starwort (*Stellaria umbellata*) and the snowgrass (*Phippsia algida*). At these elevations the environment is not as harsh as on the summit of Mt. Logan, but interpolating between the Mt. Logan weather records and those obtained at 2652 meters on the Icefield Ranges suggests that mean growing season air temperatures may be several degrees below freezing, and extreme minima may be close to -20°C.

In the relatively biologically depauperate ranges of this area of North America, the tree line is solely composed of white spruce (*Picea glauca*) and occurs at about 1200 meters. Above this level the alpine zone is dominated by a sedge, *Carex microchaeta*, and a prostrate willow, polar willow (*Salix pseudopolaris*), except for the driest sites which are dominated by a mountain dryad, *Dryas octopetala*, and a different prostrate willow, netleaf willow (*Salix reticulata*). The plants growing at the highest altitudes on nunataks are found at about 2640 meters, well above the snow line; in fact, plants have even been found at 2800 meters. On the face of it, this elevation seems relatively low, considering that vegetation in the Colorado Rockies easily reaches 4300 meters. But high latitude magnifies the biological effects of high elevation, and an increase of a few hundred meters in the arctic or subarctic regions might have the same overall impact on the vegetation as an increase of several thousand meters closer to the equator.

In these subarctic mountain ranges we find many new genera that were not present in the Arctic, but we also find a few old friends, like the moss campion, a variety of the arctic cinquefoil, and close relatives of the arctic dryad. The state of dryad nomenclature is in fact quite confusing, since the three distinct species seem to hybridize rather liberally, and it is easy to find an individual that displays intermediate traits. Other relatives include carices, fescues, whitlow-grasses (*Draba* spp.), chickweeds (*Stellaria* spp.), and a completely different set of saxifrages (*Saxifraga* spp.).

A similar set of alpine communities can be said to extend the whole length of the Rocky Mountains, to their southern limit. The southernmost mountains with a similar assemblage of species are the San Francisco Mountain in Arizona and the White Mountains of New Mexico. These two areas were some of the most southerly outposts of the Wisconsin Glaciation. The San Francisco Peaks (as the locals call it) is located just north of Flagstaff, Arizona and is a 12-million-year-old stratovolcano. Its summit was first botanically explored by H. C. Merriam in 1889, and it was from these observations that he devised his now outdated, but still historically important, concept of climate zones. This mountain receives over 76 cm of yearly precipitation close to the summit; it is located at 30°N latitude, and its highest peak is 3650 m. The plants in the inner crater of this volcano are typical Rocky Mountain alpines, with the added twist that several of them are at their southernmost distributional limit. These include the long-stalked starwort (*Stellaria longipes*), the boreal sandwort (*Minuartia rubella*), and the alpine avens (*Geum rossi*), which is the most common plant in the crater. In many parts of the crater there is a complete cover of vascular plants on the thin soil derived from pyroxene dacite and andesite.

The physiological and behavioral adaptations of animals to extremes of temperature have been thoroughly studied, especially in arthropods. There are several reasons for this preoccupation with arthropods. They are the most visible members of the fauna of high mountains. Also, because of their more manageable size (in relation to other inhabitants of harsh regions) they are often chosen for more thorough laboratory investigations. Undeniably, their astounding variety of forms also attracts investigators.

The inhabitants of the ice realm of high mountains consist mostly of several orders of insects, springtails, and mites (Class Arachnida: Order Acari). The ancient insect orders that inhabit the aeolian zones of temperate mountain ranges include the bristletails (Archeognatha), the mayflies (Ephemeroptera), and the rock crawlers (Grylloblattodea). These insects are morphologically similar to fossils 300 million years old. The grylloblattids or rock crawlers have characteristics of both the crickets (family Gryllidae) and the cockroaches (family Blattidae), which suggests that they diverged along their own unique evolutionary path before the Orthoptera (the order to which crickets belong). The grylloblattids now have a highly disjunct distribution; they are found in the high mountains of western North America, Siberia, and Japan. This disjunct distribution suggests a much more wide-ranging past.

The grylloblattids are extremely sensitive to elevated temperatures and succumb quickly if they are picked up with bare hands. For some of these insects the optimum temperature is near 0°C. For instance, the species of *Grylloblatta* (the only North American genus) have an optimum of 1°C and actively seek this temperature. These insects die if their body temperature climbs above 20°C even momentarily; therefore, they are always found near snow. Grylloblattids are much larger than the springtails with which they share their environment; they often reach 30 mm in length. Their size prevents them from inhabiting the same snow pores the springtails inhabit, so they are most often found under rocks, in logs, or crawling over the snow, where they hunt for whatever insect prey is available, alive or dead.

Even though the grylloblattids are not represented in the Arctic, other insect groups are represented both in high mountains and in the Arctic. A common crane fly in the alpine realm is the snow-fly (*Chionea* spp.) of the mountains of Europe and North America. It is often active on the surface of the snow when the temperature is -10°C. In the summer the snow-fly takes refuge from the intense sun in the burrows of wasps and small mammals, where its larvae live.

Winter Drought on Temperate Mountains

On the mountain ranges found along and east of the North American continental divide, we encounter a conspicuous winter phenomenon involving both water availability and cold temperatures. Here, desiccating air and mostly frozen soil cause a condition known as red belt. This term comes from the browning of the needles of conifers growing at midelevations on the southern or western slopes of mountain ranges. Even though the term red belt applies only to conifers, the physiological drought that is its root cause applies to all the plants growing on the mountain along with the trees. Dry air can be made in place or it can be imported with chinook winds. When it is made in place, bright winter sunshine accompanied by still air increases the temperature without increasing the moisture content of the air, and enhanced desiccation is the result. When chinooks are involved, they bring with them air that has been wrung of its moisture and in the same stroke warmed, so that its desiccating power is doubly enhanced. Under both of these conditions, the air temperature can soar on sunny winter days while the ground temperature remains close to freezing. If a plant is able to freely transpire, it can keep the temperature of its leaves within tolerable levels, and dry air is not a threat. If a plant cannot transpire freely, either because of dry soil or low soil temperatures (which inhibit water uptake by roots), the plant suffers on two fronts. Without the cooling effects of transpiration, insolation can cause the plant's tissues to reach dangerous temperatures. In order to keep cool, a plant may be forced to transpire at a rate that is too rapid for its roots to keep up. Of course, when the soil water is frozen, roots are not able to take up water, but even during times of low but nonfreezing temperatures, water uptake is slowed. One temporary solution to this bottleneck is to store water in the tissues of the plants, and this adaptation is found in some conifers such as the Douglas fir (*Pseudotsuga menziesii*), the limber pine (*Pinus flexilis*), and the ponderosa pine (*Pinus ponderosa*). A typical Douglas fir 80 m tall may store 3000 transpirable liters of water in its sapwood, but even this massive storage often is not enough to save the tree from red-belt damage.

To really understand how red-belt damage occurs, we must first understand the nuances of water movement in trees. The taller a plant is, the more energy it takes to deliver water to its leaves. Part of this job is accomplished at the level of the roots, and part of it is accomplished at the level of the leaves. At the level of the roots, water can be pumped against a concentration gradient; this process requires expenditure of energy by the plant, and only some plants are capable of this feat. At the level of the leaves, evaporation of water translates into a pull on the water columns in the tracheids of the tree's xylem (the water-conducting pipes). These tracheids are so small that the cohesive force of water (which gives water its surface tension) allows the whole column to be pulled toward the leaves. If the root system is not functioning at a high enough capacity to match the rate of evaporation at the surface of the leaves, the water column becomes stretched beyond the breaking point, and the flow of water to the leaves stops. A transpirable store of water affords some protection against this event. If internal stores are transpired away, and root function is still compromised by cold soils, water loss can lead to cavitation—the blockage by gas bubbles of the water-conducting tracheids —a condition about as life threatening to a plant as a case of the bends is to a human diver. Death of the leaves soon follows, and the whole drama becomes evident even to the casual observer as a band of red on the side of the mountain. Some trees have mechanisms to deal with this contingency. Limber pine has a greater ability to extract water from cold soil and greater control over its stomata than other species growing in the red-belt areas. With these abilities, limber pine is able to grow on some of the coldest, and at the same time driest, mountain ranges of the western United States.

If water storage and tight stomatal control are prudent strategies for survival in the harsh mountain environment, then the cacti, that uniquely American family of plants, should be able to grow and prosper in such habitats. They are, after all, water hoarders par excellence and have taken stomatal control to the state of a fine art. Popular conception is that cacti are too frost-sensitive and would be harmed by winter frosts. However, it turns out that some cacti, especially the opuntias (prickly pears and chollas), are quite cold hardy and can easily withstand mountain winters. For instance, one species of prickly pear, *Opuntia fragilis*, is documented as far north as 56°N latitude (the same latitude as Moscow), making it the northernmost cactus species. It is so common in the dry inland valleys of British Columbia that Simon Fraser—one of the early explorers of the area—commented on a "thistle" that could penetrate boots with its thorns and made travel painful and difficult. The only such "thistle" in the area is the prickly pear, normally found hiding among the bunch grasses. These cacti are in fact members of the understory community in some of the red-belt areas we have been exploring. For cacti to come into their own, however, competition for light from taller plants must be removed. We will discuss such an environment when we reach the tropical mountains.

Ecology of the High Himalayas

The Himalayas is not a single chain of mountains but a series of chains; all were born as the Indian subcontinent crashed into and crumpled the Asian mainland 40 million years ago. Starting on the highest series of peaks (which includes Mt. Everest) and moving northward, we encounter smaller mountain ranges that gradually coalesce into the great Tibetan Plateau. The monsoonal moisture that comes in from the Bay of Bengal gets deposited as snow on the southern slope of the first range, leaving little moisture for the ranges that are hidden by these giant mountains. This orographic or rain-shadow effect is what makes parts of western Tibet a desert with less than 100 mm of precipitation a year. One hundred millimeters of precipitation is

meager by any standard, and more so in this case, because of the more intense and more direct sunshine (caused by the high altitude and the relatively low latitude respectively). Therefore, sun-exposed areas can be extremely dry. We will first explore the high Himalayas and then move on to Tibet and the Chang Tang.

On the north side of the Nepalese Himalayas, on the Chinese border, lies Makalu (27.54°N latitude). One of the few glimpses of ultrahigh-altitude life comes from scientific expeditions to this mountain in 1956, 1960, and 1971. This mountain rises 8481 meters (by comparison, Everest rises over 8848 meters) and, like the other mountains of this range, the northern slope is drier and has less snow cover than the southern slope. A seven-day weather record taken in April 1971, at an elevation of 5800 meters, shows that the temperature at 5 cm above the ground fluctuated from a low of -19.2°C to a high of 20.8°C, with a mean value of -5.7°C.

High up on the northern slopes of the Himalayas we find flies of the family Anthomyiidae. We have already met this family in the High Arctic, where their special talents seem to be spurring a new bout of evolutionary expansion. These flies are common on the northern slopes of Makalu, at altitudes above 5500 meters. The larvae of most of the species in this family live in dung or in decaying plant material. At these high altitudes there are hardly any plants, let alone decaying plant material, and the only animals that habitually make these high peaks their home are snow partridges (*Lerwa lerwa*), which have been known to nest as high as 5800 meters. It may be that these flies are commensals in the nests of these birds, feeding on excrement and other nutritious substances.

At these altitudes in the Himalayas life is very harsh. Freezing temperatures can happen any night of the year, and there is little precipitation (less than 38 cm a year). Because this alpine area is relatively close to the equator, the sun beats down mercilessly. What little snow does fall evaporates without melting (a process known as sublimation). Because the air is dry, dew does not normally condense. Above 5500 meters, the very sparse plant cover is confined to sandy areas and beneath boulders, places where moisture has a chance of being trapped before it evaporates. This transition zone, where even the hardiest plants must seek shelter, has been termed by L.W. Swan the alto-alpine or rock-base zone.

From a distance or from an orbiting satellite, only the large-scale altitudinal limit to plant life can be observed: green below, dark barren rock above; but upon closer inspection, a more indistinct limit higher up the mountain is always found. On Makalu, plants maintain a tenuous hold up to 6100 meters, where grasses and sedges grow in areas where subsurface water is available, and other plants including sandworts (*Arenaria* spp.), edelweiss (*Leontopodium* spp.), cottony saussurea (*Saussurea gossypiphora*), a gentian, *Gentiana urnula*, and a parrya, *Parrya languinosa*, are confined to rock bases. At 6136 meters the last flowering plant found by L. W. Swan, hidden under rocks, was the starwort *Stellaria decumbens*; he found no other plants around it, not even lichens. At this altitude, snow and ice sublime rapidly under the influence of the unfiltered solar radiation; only areas with large rocks or sand will be able to collect water under the surface and keep it from the drying rays. Life is relegated to sheltered cracks and crevasses as the snow line is approached.

Along the Himalayas of Uttar Pradesh, the families that dominate the high alpine realm are the same ones present on Makalu, but the protagonists change because of changes in water availability and substrate type. In these other mountains, the crucifers such as *Alyssum*, *Cardamine*, *Chorispora*, *Christolea*, *Cochlearia*, *Draba*, and *Parrya* survive above 5400 meters. (The majority of these genera are also found in the Arctic.) In this region the starwort, which was king of the mountain at Makalu, only grows to an altitude of 5000 meters, and the title has been claimed by the crucifer *Christolea himalayensis*, growing at 6350 meters on Mt. Kamet.

In the protected rock-base sites, a variety of arthropods can be found. The herbivorous mites are preyed upon by carnivorous mites, which also prey on the springtails. Larger predators include small centipedes and jumping spiders (family Salticidae). The jumping spiders are the largest and the most active predators of this community. These spiders were first observed during the early expeditions to Mt Everest, at 6705 meters, and they are relatively common on nearby mountains as well. The nature of these spiders' diet is still somewhat of a puzzle, but since they are diurnal sight hunters, their quarry might be anything smaller than themselves, and most likely consists of springtails and flies. Lower than 5500 meters, the wolf spiders become dominant, and the jumping spiders almost vanish. The reason for this transition is not known, but wolf spiders are probably restricted to lower altitudes because of their lower cold tolerance, and jumping spiders probably become wolf spiders' prey when the two occur together.

Few vertebrate herbivores make the upper alpine zone their home. The snow partridge sometimes nests above 5500 meters and feeds on roots and leaves around its nesting site. Mice and pikas (small lagomorphs related to rabbits but that look like mice) are the only other vertebrate herbivores that sometimes reside there. Because of the scarcity of resident prey, resident predators are scarce as well. Only the Tibetan weasel (*Mustela altaica*), which feeds on snow partridges' eggs, may be able to make a meager living above 5500 meters. Unlike its high arctic relative the ermine (*Mustela erminea*), this weasel does not have the protection of lemming burrows to fend off the cold and must do with any kind of rock crevice.

Past the rock-base realm we enter what L. W. Swan christened the aeolian zone. Here plants are not in evidence, and the communities found at this altitude must subsist on wind-blown debris brought up from the more vegetated zones below. This is the home of the bristletails of the family Machilidae that live on the rubble covering the ice of glaciers. Other arthropods such as phyllopod crustaceans and midges thrive in the pools of water that collect on the surface of glaciers. These pools act as traps for windblown debris and provide a congenial environment for those animals that can withstand the frigid conditions. Glacier streams that trickle through the cracks in the ice and emerge undaunted and full of energy at the base of the glacier also provide suitable environments for life. In fact, these streams support large populations of stoneflies and mayflies. Other insects live in the dry interstices between the snow of permanent snow fields and glaciers. This is a much more stable environment than the surface of the snow. One of the most highly studied insects in this environment is the springtail *Isotoma*. The members of this genus are black in color and just over a millimeter long; sometimes, in the high Alps, they spread out over the snow and create a "black snow" effect. This springtail genus is represented in several mountain ranges including the Alps of Europe and the Himalayas. Under the snow these animals seek temperatures of 0 to -4°C during the summer. In the winter they go down deeper and are active at temperatures of -3 to -5°C. (Most of the snow fauna avoid temperatures higher than 10°C.)

The machilid bristletails are less well represented in the alpine fauna, but they are more conspicuous. In the Eastern European Alps, the jumping bristletail *Machilis fuscistylis* lives on bare rock surfaces and rock surfaces covered by lichens, where it presumably scrapes the rocks to obtain sustenance. Its mottled dark coloration makes it almost invisible against the dark rock. In a similar environment, but at 5330 to 5790 meters in the Himalayas, another machilid, a glacier flea, genus *Machilanus*, probably feeds on the airborne particles brought up the mountain by the winds. Very little is known about these high-altitude insects, and it is remarkable to observe the preponderance of primitive insects and proto-insects in these habitats. Or as L. W. Swan wrote:

> Perhaps there is a distant analogy between the highest altitudes and the sterile land of the Silurian period, when air-breathing arthropods evolved from their aquatic ancestors. These same primitive insect types may have been among the vanguard of animals on land, and it may be that the first land animals fed on the wind-blown debris that accumulated among the barren rocks of the world beyond the fringe of shore plants. (Swan, 1961)

Ecology of the Chang Tang

In the high plateaus between the Himalayas and the Kunlun mountains there are large expanses of grasslands and rolling hills that rise several hundred meters above the general elevation of the 4900-meter-high plateaus. This region is known in Tibetan as the Chang Tang. These steppes support a variety of ungulates, lagomorphs, and rodents. As we go north and west from the central Tibetan plateau, precipitation drops until, close to the Kunlun mountains, it is less than 100 mm a year. This region, with only ten percent plant cover, approaches our definition of an extreme desert. What distinguishes this part of the Chang Tang from so many other areas with scant rainfall is the altitude. These extensive high plateaus are found only in the heart of Asia and on the spine of the Andes. In both these areas the plateaus are ringed by a crown of lofty peaks that block life-giving rain from reaching the plateau regions. The plants that have evolved in these two remote regions are of different ancestry, but share the same vicissitudes. In the northwestern Tibetan plateau, north of 35°N, large tracts of land are almost devoid of vegetation. The dominant plants of this high desert are a coarse sedge, *Carex moorcroftii*, a dwarf shrubby chenopod, *Ceratoides compacta*, and a needlegrass, *Stipa glareosa*. The latter two plants have been shown to be resistant to temperatures of -16 or -17°C, or even lower, during the growing season.

Few mammals can survive such extremes of both cold and drought. The primary ungulates of this region are the chiru (*Pantholops hodgsoni*) and the yak (*Bos grunniens*). The chiru or Tibetan antelope is not an antelope at all, but is related instead to sheep and goats, not fitting happily into either group. These animals have some goat traits such as glands in the groin and at the base of the hoof, but the females look remarkably sheeplike at a distance. Chirus are somewhat migratory, and their migrations are remarkable in that, like some other denizens of extreme environments, the females move to extremely inhospitable areas to give birth—in the case of the chirus, from late June to early July. The previous year's female offspring accompany their mothers on this arduous journey, but male chirus only go partway, and in typical male fashion, never experience the ordeal in its entirety. The places chosen are probably barren uplands above 5200 meters. (No report has emerged of the exact location, as this is truly an uninhabited part of the planet.) In these areas heavy snows can fall even in early July (with temperatures falling to -20°C), and the does run the risk of starvation or of their infants' freezing to death. The females do not loiter in the barren lands, but soon embark on the journey back to more lush pastures where snow does not accumulate in midsummer. The total journey may have taken one month and spanned 600 km. As mentioned above, yak too live in this region. The few remaining wild yak herds roam these desert areas where hunting pressures are less, and here they range over the rolling hills up to the limit of plant growth.

Two other little-known ungulates frequent those high elevations. They are the Tibetan argali (*Ovis ammon hodgsoni*) and the bharal or blue sheep (*Pseudois nayaur*). The Tibetan argali is confined to the Tibetan plateau where it prefers high slopes. Very little is known about it, and the chance to find out more seems to be quickly slipping away, for its population seems to be in rapid decline. In 1903, C. Rawlings described argali in large numbers grazing in the ravines of the Aru Range in the Chang Tang; only a few can be seen nowadays over the whole of the northern plateau. This argali is only one of a number of subspecies of a much more widespread species that, like many other organisms with a predilection for mountains, has diverged into several reproductively isolated populations. In central Asia, other mountain ranges, such as the Altai in Mongolia and the Pamirs, have their own subspecies of argali, each adapted to the particular nuances of its home range. The bharal is a different story; it is not confined to the Tibetan plateau and in fact, *bharal* is an Indian name for this animal. It is found throughout the Himalayan complex, even crossing south into Ladak. It frequents more precipitous terrain than the argali and is therefore very difficult to study. It is known that it ranges over high mountain peaks, feeding on a variety of plants including some graminoids such as sedges of the genus *Kobresia* and grasses of the genus *Poa.* But it prefers forbs, including one of the Himalayan edelweiss, *Leontopodium pussilum*, milk-vetches (*Oxytropis* spp.), and a cinquefoil, *Potentilla bifurca.* These plants compose only about one third of the available ground cover close to the altitudinal limit of vascular plants, but constitute over 80% of the bharals' diet, which shows that these animals have definite preferences.

The least palatable plants of the high plateaus are the coarse sedges (mostly *Carex moorcroftii*) and the dwarf shrubs (mostly *Ceratoides compacta*). These are, however, the most resilient vascular plants, and almost the only ones found north of the Chang Tang on the way to the Takla Makan, one of the most desolate places on Earth. Only yaks and chirus browse the stems of the *Ceratoides* and the sedges as a significant part of their diet. There are only a few wild yak herds left in the Chang Tang. They are mostly relegated to the northwest, to the barren lands where they can avoid human beings by putting vast swaths of land between them and the nomadic herdsmen that inhabit the plateau. In this way, wild yak herds are limited in their choice of browse. They seem to choose the subshrub over the sedge, in direct contrast with the domesticated yak herds in the Aru basin, which shun the shrub completely. Maybe the wild yak herds, with their communal memories of places to go and places to avoid, have a few secret patches of tasty *Ceratoides* to draw upon. The domestic yak, on the other hand, has much more limited choices and can only say yea or nay to the forage in front of its nose. The dwarf shrub *Ceratoides compacta* has a close relative in the American salt deserts: winterfat (*Ceratoides lanata*), so called because it makes excellent winter forage. These two *Ceratoides* species are similar in their physiological limitations. The American species can grow in areas up to 3100 meters in elevation and receiving as little as 100 mm of annual precipitation. The lower elevation limit in the American species may be due to the more severe winter cold on American mountain ranges. A glimpse into the genetic variations found within this genus can be had in the cold dry lowlands of southwestern Yukon, where there is a disjunct population of winterfat that may have a greater cold hardiness than the populations closest to it, located in eastern Washington state. Other evidence of the genetic adaptability of *Ceratoides* lies in the pronounced variations detected in the germination temperature of seeds from different geographic populations of winterfat, even when these populations are separated by only a few hundred kilometers.

The Tropical Alpine Environment

The puna, the high arid plateau of the central Andes, is too cold and dry for Andean subalpine trees. On this arid plateau, opuntias and their relatives have unhindered access to light and have developed several remarkable forms. One of these—formerly *Opuntia floccosa* but now split off into the genus *Tephrocactus*—grows in the puna at 4570 meters on grassy high mountain slopes without any trees, making it the highest-altitude cactus in the world. When viewed from a distance, it looks for all the world like patches of snow, or as O. F. Cook wrote:

> Many exposed slopes on the bleak plateaus of the high Andes are dotted with clumps of pure white cacti that look from a distance like small masses of snow. On closer view, the shaggy white hair of these cacti make them appear like small sheep or poodle-dogs, or like reduced caricatures of the denizens of the arctic regions. We are so accustomed to think of cacti primarily as desert plants, peculiarly adapted to hot, dry deserts, that they seem distinctly out of place on the cold plateaus of the high Andes... (Cook, 1917)

These wolves in sheep's clothing (for they will show their teeth if you brush against them) are not the only plants to adopt this animal-like woolliness. The woolliness that made these plants so strange to foreign eyes is an often-used adaptation of some very dissimilar plants.

Because of the closeness to the equator, this altitude in the high altiplano is not as cold as similar elevations in the Northern Hemisphere, but these tropical highlands suffer large fluctuations in temperature. The average monthly minimum temperature is below freezing for six months of the year, with extremes of 35°C and -18°C. (In the southern Andes, temperatures are even more extreme, and yet other genera of cacti can grow there up to the snow line.)

In the case of opuntias, the mechanisms that allow them to withstand winter cold (down to -70°C in one test) involve desiccation and concentration of the internal solutes to such an extent that most internal functions cease. This process is cued to the shortening day length; this is the only way a plant can assure itself enough lead time to complete winter hardening. By the time winter sets in, the opuntias have dehydrated and have become wrinkled bags holding a syrupy solution. Since recovery is a long process, this mechanism of protection is fairly drastic and not to be entered into lightly.

The mechanisms that opuntias, as well as other high-altitude cacti, use to withstand frosts during the growing season are unknown. It may be that the thermal time constant of the colonies these plants form is long enough, so that the temperature inside plant tissues never dips below zero for very long. Opuntias are a rapidly evolving group; what kind of frost tolerance they will develop next no one can tell.

Cacti share the bleakest, driest puna habitats with cushion plants called yaretas (*Azorella* spp.) that form unique rock-like features of the landscape. In a landscape where they are the most productive vegetation, these plants serve as sources of fuel for the human inhabitants of these harsh regions. Their value as fuel and their slow growth rate have led to their disappearance over large areas of puna, to the point that now they are easily observed only in the national parks of Peru and Chile. These plants embody another facet in the history of mountain colonization, for unlike the other colonizers we have talked about, they came from the south. Even now there are several species of *Azorella* in Southern Hemisphere islands, even as far away as Macquarie Island, south of New Zealand. The mound-like habit of these plants was most likely evolved as a solution to heat loss in cold, moist, and windy environments, but it has served them equally well in the cold, dry puna.

Little is known about the adaptations of plants in the southern puna where cacti, dwarf composite shrubs and grasses (mainly *Stipa jehu*) reign up to the limit of plant life at 4600 to 4800 meters. However, in high-altitude habitats closer to the equator, like in the northern Andes and in the high mountains of Africa, certain plants have embarked on a course of amazingly parallel evolution, producing forms that most closely resemble the yuccas and aloes of American and African deserts. When taken together, these adaptations, both morphological and physiological, constitute an adaptation syndrome that is unique to environments with extreme daily temperature fluctuations at high altitude.

These alpine plants go by several names including giant rosette plants, caulescent rosettes, and megaphytes. Their morphological similarities are striking, but their physiology might also be quite similar. On Mt. Kilimanjaro and Mt. Kenya, the megaphytes are of two types: the giant groundsels now placed in the genus *Senecio*, and the giant lobelias (*Lobelia* spp.), the same genus as the cardinal flower, or scarlet lobelia, of eastern North America. On the páramo of the northern Andes, the megaphytes belong to the following genera: *Lupinus* (Fabaceae), *Puya* (Bromeliaceae), *Valeriana* (Valerianaceae), *Draba* (Brassicaceae) and *Espeletia* (Asteraceae). In the central Andes, where the puna is comparatively drier and the weather more seasonal than in the Venezuelan páramo, the bromeliads gain their greatest stature. Here, the gigantic *Puya raimondii* reaches 10 meters tall when in flower. The overall effect is very similar to that seen in the Baja California desert mountains, where yuccas and nolinas take on megaphyte airs, but with less panache. Even in the high valleys of Mt. Kenya and Mt. Kilimanjaro, where the giant groundsels and lobelias grow, they are at the base of a food chain that includes small herbivores like hyraxes and large ones like elephants.

The genus *Draba* that is present in the arctic and in the temperate mountain ranges is present in the páramo also. In fact, while eight perennial species of *Draba* occur in Peary Land in northern Greenland, seven perennial species of *Draba* occur in the Venezuelan páramo (however, there are no species common to both locations). In the páramo of Venezuela the plant growing at the highest altitudes is the "miniature" caulescent rosette *Draba chionophila* ("miniature" compared to neighboring megaphytes). The drabas that grow in the Andes are morphologically

diverse, unlike their northern temperate counterparts that are mostly small ground-hugging rosettes. The longer growing season in the páramo allows some of the South American drabas to produce large-leaved rosettes, and the lack of a distinctly cold winter allows these leaves to be evergreen. Some species occur as shrubs; others, like *D. chionophila*, are miniature caulescent rosettes. Their morphological variety belies their genetic similarity. All the Andean drabas have been shown to have a complement of 24 chromosomes—an unlikely coincidence if these species had polyphyletic origins.

Draba chionophila has been more intensively studied than most, since it is the highest-altitude vascular plant in the Venezuelan páramo. It looks like a miniature version of the frailejones (*Espeletia* spp.) with which it shares the páramo, but its stem is only 5 cm tall (up to 20 cm in some specimens), and the leaves can be 5 to 10 cm in length. Unlike the *Espeletia* growing farther down the mountain sides, *D. chionophila* is a freeze-tolerant species. It is found above 4500 meters; at that altitude the páramo is devoid of almost all plants, possibly because of the increased daily temperature fluctuations at the surface of the ground. Night frosts here can dip below -10°C, and *D. chionophila's* strategy to deal with this threat is to freeze when the temperature drops below -5°C. Farther up the mountain sides, even it is overwhelmed by frosts that dip lower than -14°C, its absolute low-temperature limit.

The local environment of these megaphytes is produced by a combination of high altitude and relatively low latitude. These two factors conspire to produce a relatively short day length with high temperatures, and a relatively long cold night when freezing is the norm rather than the exception. On some tall mountains, the precipitation maximum occurs at low to midelevations. Upwards from these relatively wet altitudes, the decreased precipitation, together with the decreased air pressure, makes liquid water a fleeting commodity. At low elevations in the early morning, when plants are first hit by the rays of the sun, a moist atmosphere allows the plants to ease into the work-a-day world of photosynthesis. At high elevations the air holds much less moisture, and without this protective moisture, plants are propelled directly into the harsh desiccating reality of heat and sun.

The low water availability is compounded by soils that freeze during the night and might not thaw out for several hours after sunrise. This effective drought state comes at a time in the early morning when leaf tissues are being heated by the sun and photosynthesis and concomitant transpiration beckon. (This predicament is not exactly the same as that encountered in red-belt areas, for there the trees are dormant when red-belt conditions arise.) There are several possible ways a plant could handle such a Gordian knot.

Nighttime CO_2 uptake would solve part of the problem faced by the megaphytes, but they may still need to open their stomata in the early morning, in order to cool down as the sun beats down relentlessly on their leaves. This overheating problem might be solved by judicious positioning of the leaves in the early morning, and as we shall see, some megaphytes move their leaves to the daily rhythm of freezing and thawing. The other solution, and the one that megaphytes seem to prefer, is to store enough water; large rosettes with thick stems can store sufficient water within their tissues to supply their transpiration demands for several hours. Of course, this strategy depends on there being enough soil water to replenish this internal store sometime later in the day. This water-storing solution allows these plants to utilize the early morning sunlight.

Carbon dioxide is one of the raw materials (along with water and sunlight) plants use to make sugars in the process known as photosynthesis. As we saw in chapter 2, some plants have developed biochemical pathways that allow them to separate the uptake of carbon dioxide from the rest of the reactions that make up photosynthesis. This feat has allowed them to open their stomata and let in carbon dioxide at night, when the relative humidity is high and transpiration is therefore low. The carbon dioxide is temporarily stored in chemical compounds that will be used to make sugars during the following day's photosynthesis. This approach, very popular among desert plants, would allow the megaphytes to keep their stomata tightly closed while they wait for the soil to thaw in the early morning. There is a fly in the ointment, however. For this approach to work, the night frosts must not inhibit the uptake of CO_2. This approach may have been adopted by some of the plants that share the puna with the megaphytes, but no megaphyte has yet been proven to use it except in times of drought.

Even though there is little seasonal variation in average temperature, the daily temperature variation of the páramo is severe (although not as severe as in the puna farther south). Within 5 cm off the ground, temperatures might fluctuate from a high of 20°C to a low just before sunrise of -10°C. The temperature fluctuations at some distance above the surface are less severe, and this may have been a powerful force in the evolution of the megaphytes. To protect their tender terminal bud, many megaphytes have the ability to close the rosette of old leaves over the bud—a feat known as nyctinasty—thus protecting the bud with a thick layer of mature leaves. Nyctinasty has evolved in several unrelated megaphytes—*Senecio*, *Lobelia*, and *Espeletia*—a testimony to the force exerted by similar environments on different lineages. The rosettes of some species are shaped like a parabolic dish and are able to hold several liters of liquid. It is not simply rainwater that has collected in the cup; if the liquid is removed, the plant will secrete another bowlful. This liquid comes in handy as a temperature buffer. Rosettes full of liquid are found in *Senecio*, *Lobelia*, and *Valeriana*. In all these cases the liquid is slimy to the touch, and, at least in one of the lobelias, the sliminess can be attributed to pectin produced by the plant.

When water freezes, it releases heat. This might seem strange until one thinks of the reverse process. Ice in a glass of ice water is not in equilibrium with its surroundings; as the ice melts, water at 0°C is released. The ice-water system does not increase in temperature until all of the ice is gone. Yet heat was absorbed from the surroundings to melt the ice. The heat that was absorbed from the surroundings is now contained in the 0°C water, and when the water is refrozen, it will release the heat anew. Water is then an extremely good heat-storage material right at 0°C (in fact there is none better). The goal for the megaphyte is to keep the apical (or terminal) bud as warm as possible for as long as possible. In order to do this, the liquid used must be not only a good heat-storage medium, but also a good insulator. The bad news is that water has a relatively high thermal conductivity; fortunately, part of this conductivity is due to convection and convection can be slowed, even stopped. A solution of pectin and water is very viscous; heating such a solution does not generate the usual convective cells one sees

in a boiling pot on the stove. In other words, the increased viscosity inhibits convection and enhances the insulating properties of water. Another adaptation common to many megaphytes is that of closely appressed stem leaves. Since a leaf can provide insulation whether alive or dead, megaphytes retain the old leaves on the stem for many years.

Dry Mountain Tops

We have already seen that Makalu encompasses limiting conditions for life, but the environmental gradients that lead to these limiting conditions do so at a leisurely pace, and life peters out over hundreds of meters of increasing altitude. The interplay between water and temperature is hard to discern at Makalu, but there are other drier high mountain areas where drought rapidly outpaces cold as the limiting factor for life. As explained earlier in this chapter, the puna is one of those areas, and other such areas can be found on isolated mountains in the tropics. Three of these mountains are Pico de Orizaba (5675 meters), Mauna Kea (4207 m), and Mt. Kilimanjaro (5896 m). The first is a high volcanic peak 125 kilometers west of Veracruz, Mexico. Unlike the other two, this mountain (the third highest mountain in North America) does not have any large caulescent rosette (megaphyte) species, and in fact, such species are completely absent from all the high Mexican volcanoes. The reasons for this are not entirely clear, since closely related species are found there, but with the usual North American morphology. The alpine areas on Pico de Orizaba are populated by the tussock grasses *Festuca tolucensis* and *Calamagrostis tolucensis*, which reach their upper elevational limit at around 4800 meters. Above this elevation a few lichens and mosses survive for a few hundred meters, until they too are vanquished by the mountain.

On Mauna Kea, in Hawaii, yearly precipitation close to the summit is even lower than at equivalent elevations on Pico de Orizaba, amounting to only 146.5 mm at the summit, and the average monthly temperature fluctuates 4°C instead of 2.5°C. On Mauna Kea we find scattered caulescent rosette forms. The most distinctive of these is the silversword (*Argyroxiphium sandwicense*), which forms a tight ball of sword-shaped silvery leaves, and when in bloom makes an astounding sight. This genus, along with two other tarweed genera on the Hawaiian Islands, is thought to have evolved from a single North American ancestor, and its closest relatives outside of the Hawaiian Islands can be found in California. The Californian tarweeds are prostrate annuals and have none of the grandeur of their tropical relatives. The silverswords are monocarpic—long-lived plants that flower only once, producing a large inflorescence, and then die. This habit means that these plants must rely heavily on seed propagation.

(This is the case for many of the other giant rosette plants, including the puyas and the lobelias.) The highest altitude at which the silverswords thrive on Mauna Kea receives a fair amount of moisture (equivalent to that of the alpine zone of Pico de Orizaba). Higher up the mountain, vegetation is scant, and only a fern and a grass grow within a hundred meters of the summit.

On Mt. Kilimanjaro the vegetation slowly disappears above 4800 meters, and as the summits are reached, the mountain becomes a true desert with only 15 mm of precipitation a year. This precipitation value makes Mt. Kilimanjaro the driest high mountain on Earth. Even here, however, there are special microsites that allow vascular plants to survive. Mt. Kilimanjaro is an inactive saddle-shaped volcano, but it does have active fumaroles near at least one of its summits. On the higher summit called Kibo (at 5896 meters, Kibo is the highest peak in Africa), some of these fumaroles are located around the Reusch Crater found at 5800 meters. They actively spew out steam and sulfurous vapors. Near the fumaroles several vascular plants have been found. They include *Senecio telekii*, *Senecio meyeri-johannis*, the composite shrub *Helichrysum newii*, and the alpine rock cress *Arabis alpina*. Apparently these plants can survive due to the condensing steam that coats the rocks with a film of water, thus producing a more favorable water and temperature regime. Do not be fooled; this is still a harsh environment, as can be readily verified by the presence of only two small lichens and by the low number of individual vascular plants. These small populations are at the whim of relic volcanic activity, which waxes and wanes over a period of decades, sending the populations into spasms of extinction and recolonization.

The last of our volcanic oases is perhaps the most intriguing, since it has many of the same features as Mt. Kilimanjaro, but there is not a single vascular plant to be found anywhere. This is the summit of the Socompa volcano in the Argentinean Andes. This andesitic volcano reaches a height of 6060 meters, and at 6000 meters it has an active fumarole that unbelievably nurtures a closed vegetative community made up of mosses, liverworts, and lichens. Associated with these are mites, flies, beetles, moths, earwigs, furtive rodents, and a song bird.

11 The Environments of Mars

A Few Preliminaries

This chapter speculates about the possible forms of life that might have left traces on Mars or might still be timidly hiding, waiting for a new day when the waters will return and life will be renewed. As of the first edition of this book, the Martian surface has been successfully sampled by only two spacecrafts. The estimates of climatological parameters are based in large part on the limited data of the Mariner and Viking spacecrafts, and it would be a mistake to suppose that what these spacecrafts recorded for a period of a few years is the norm rather than the exception. The unpredictability of the Martian climate should never be underestimated. Since the Viking spacecrafts landed on the red planet twenty years ago, the mean surface temperature has dropped by 20°C. New data will be gathered by spacecrafts that are now on their way to Mars, and I am looking forward to expanding this chapter in the next edition. If life exists on Mars, the discovery will herald a new age of exploration and hopefully understanding. If it doesn't, then maybe we can give life a push, and sit back and enjoy the ride.

Knowledge of the conditions on the surface of Mars comes from the Viking missions and from remotely sensed images. The data are scant, but some educated guesses can be made about the constitution and composition of the surface, the presence of water, the temperature fluctuation over a day and also over the seasons.

First, let us look at the differences between Mars and Earth from a biological perspective. On Earth, the cycling of several biologically important elements plays an integral part in sustaining the biosphere, and in most cases, the biosphere partially mediates that cycling. I do not mean to imply that the biosphere controls its own destiny. I leave that type of speculation to the philosophers, but I simply want to emphasize that any planet with an extensive biosphere will stand out among its neighbors, even from afar. The presence of an extensive biology shifts the compositions of the atmosphere and of the surface of a planet from those which would be expected if there were only physical processes at work. That changed composition can be detected from afar by using a variety of electromagnetic sensors and techniques. Unfortunately, our present knowledge allows us to predict the outward differences between living and dead planets for only a very small class of planets. Such predictions are most accurate for worlds that are not too different from our own in terms of light, overall composition, and size.

If it were not for nitrate-reducing microorganisms, our atmosphere would have much lower concentrations of nitrogen; if it were not for photosynthetic organisms, it would have a much lower concentration of oxygen and a larger concentration of carbon dioxide. Thus the global cycling of carbon, nitrogen and oxygen, as well as that of other biologically important elements, is mediated by the biosphere. Biologically important elements such as sulfur and selenium are made into gaseous compounds by microorganisms and algae and in this form are able to travel long distances from the point of generation. Only global cycling of water and carbon dioxide are probably still active processes on Mars. These Martian cycles can be totally explained by physical processes. Thus, outwardly, Mars is not a living world, but there is still the chance that it contains cryptic life. In the following pages I will use the terms Martian plant and Martian animal to refer to hypothetical photosynthetic and nonphotosynthetic organisms respectively, without implying any other familiar traits.

The Evidence for Life

Twenty years after the Viking landers tested for life on Mars, there is still debate as to the meaning of the results. The debate centers on the Labeled Release experiment in which Martian regolith material was incubated with water and radioactively-labeled sugar for several hours. The release of radioactively-labeled carbon dioxide was supposed to be an indication of the existence of life on the planet's surface. However, the release of CO_2 was too rapid, which suggested a reaction with a soil-based chemical oxidant instead of a biologically mediated oxidation. The experiment that was supposed to corroborate the presence of life, the gas chromatography/mass spectrometry (GC/MS) experiment, did not detect any organic matter. Thus the conclusion drawn by most of the scientists involved was that the positive result of the Labeled Release experiment must have been due to chemical and not to biological action. However, a few scientists still disagree; according to G. V. Levin (one of the original designers of the Labeled Release experiment), the Labeled Release experiment can detect as few as 10 respiring cells. This experiment is sensitive enough to detect the microorganisms present in the dry valleys of Antarctica—where, as in Mars, the organic matter concentration is too low to be detected by GC/MS. Laboratory experiments suggest that the Martian surface is a highly oxidizing mix of unstable oxides of iron as well as other elements. Such a surface would certainly have produced the observed results, but it could also have masked any native microbial activity.

The evidence in favor of life on Mars has always been circumstantial, but it is becoming stronger as more Martian meteorites are examined. There is very good evidence that Mars once contained significant quantities of organic matter, since the SNC and the Allan Hills meteorites, which originated on Mars, contain organic compounds. Now, through the work of D. McKay, E. Gibson, and K. Thomas-Keprta, we may have a glimpse into early Martian life. If their evidence is corroborated, and not found to be some strange mineralogical phenomenon, then we may ask the question: Could life have survived as the conditions became progressively more severe? Current wisdom expects any surviving Martian life to be microscopic and possibly ecologically similar to the endolithic lichens found in the dry valleys of Antarctica. On the other hand, there may be protected enclaves under the surface where life may have evolved into more varied forms.

The Thought Experiment

Think of this chapter, not so much as a flight of fancy, but as a thought experiment, meant to explore the physical limitations imposed on living organisms by the environment. We must pay attention to even the most basic physical principles in order to come up with possibilities instead of with mere fiction.

In this experiment we have several tools to work with. For materials we have the first 50 meters of the regolith (the loose rock that takes the place of soil on other worlds) and the atmosphere's composition. For possible sources of energy we have the sunlight that falls on the surface of the regolith or some form of high-energy chemical compound. With sunlight may come high levels of ionizing or damaging radiation. The physical factors that will affect what we can do with these materials and energy sources are temperature, pressure, and the availability of moisture (which in turn is highly dependent on temperature and pressure). These physical factors will dictate not only what we can do, but also when we can do it, since they will change over the course of the Martian day and the Martian year.

On Mars, freezing temperatures are the most obvious limit to biological activity. Even in the tropics during the Martian summer, temperatures can fluctuate from a few degrees above freezing at midday to -40°C and below at night. Nowhere on Earth is this fluctuation as extreme as it is on Mars, but we can extrapolate from the adaptations we see on mountain tops to come up with possible Martian adaptations. The very thin atmosphere of Mars places physical limits on the availability of water and oxygen. The previous chapters have shown us the kinds of adaptations that Earth-based life has developed to deal with similar limits. Extrapolating from these adaptations to those necessary to succeed under Martian conditions is the goal of our thought experiment. In each instance we will discuss Earth-based organisms that may have parallels on Mars, and the adaptations necessary to allow the hypothetical Martian organisms to survive in their environment. We will proceed from microbial to metazoan organisms in a progression from the plausible to the implausible, not because such organisms cannot evolve, but because they may not have had time to evolve. The final section discusses some of the design parameters that must be taken into account for a Mars-based agriculture to succeed sometime in the future.

Biological Evolution on Mars

Since Mars is compositionally similar to Earth and has an ample supply of carbon, a fair assumption is that any life that may have evolved on Mars would be carbon-based. The other main chemical requirement of life is an appropriate solvent that will allow reactions to take place at a reasonable rate. On Mars this solvent would have had to be water, at least in the initial phase of life (as is the case on Earth). Large scale fluvial features first photographed by the Viking orbiters, as well as heat-flow models, suggest

One of the densest drainage networks on Mars, located at 48°S, 98°W. The picture is 150 km across. (Courtesy of C. P. McKay, NASA AMES)

that Mars may have had flowing water for several hundred million years in its early history, after the end of the planetary bombardment 3.8 Gyr ($3.8x10^9$ years) ago. During the planetary bombardment, the formed planets absorbed the remaining chunks of matter left from the accretion of the solar system, like greedy hatchlings devouring their smaller nest mates.

If Mars did have a few hundred million years of warmer and wetter conditions, this period may have lasted long enough for life to have evolved. On Earth, fossilized prokaryotic organisms have been found in sedimentary layers formed 300 million years after the end of the planetary bombardment. On the other hand, the first discernible eukaryotic fossil has been dated to around 1.4 Gyr ago or over $2x10^9$ years after the appearance of the first discernible cells. It is possible that, in the few hundred million years allowed by current thinking, life on Mars generated what we would label as prokaryotic organisms. The recent discoveries within ALH80041 bolster this argument. However, we are, to coin a word, "eukaryocentric." It is a widely held belief that all the marvelous inventions of life such as sex, eyes, and flight, could not have arisen without the clever compartmentalization found in the eukaryotic cell. On the other hand, once you have prokaryotic organisms, it is only a matter of time until partnerships are forged. Once these "consortia" become intimate enough, one partner can become an internal part of the other, and we have the first step toward a eukaryote. If these endocellular symbioses occurred, the types of adaptations that will be discussed in the following pages may have had time to appear as the planet slowly cooled and dried to its current condition. On Earth, several different kinds of eukaryotes arose, as is evident from the differences among the nuclear apparatus of dinoflagellates, most fungi, and most other eukaryotes. The evolution of Mars toward its current state of extreme cold, anoxia, and aridity may have spurred the evolution of unique eukaryotes.

Icebound Lakes

On Earth, even in the harshest regions, there are oases of life. These might take the form of lomas that intercept fog moisture as we have seen happens in coastal deserts, or they may be lakes, as we have seen happens in the dry valleys of Victoria Land. In the Arctic, lakes can be benign habitats compared with the surrounding tundra. The food chain in many such Arctic lakes is based on Richardson's pondweed (*Potamogeton richardsonii*) and several other species of pondweeds that are important sources of food for fish and waterfowl. One of the few other plants to be found in the High Arctic, from Baffin Island and southern Greenland to Ellesmere Island, is mare's-tail (*Hippuris vulgaris*, family Hippuridaceae); many times it is the only submerged flowering plant. Together with the water-milfoils (*Myriophyllum* spp., family Haloragaceae), these submerged vascular plants supply O_2 to the waters of icebound lakes. Some large lakes on the Mackenzie's Delta are well stocked with whitefish (*Coregonus* spp.) that attain large size. These fish even spend the winter in these lakes, and the Delta's Inuit use the presence of a good bottom cover of pondweeds as an indication that a lake will yield a good catch of fish.

A few freshwater lakes in the High Arctic have been extensively studied. The best known of these is Lake Hazen in northern Ellesmere Island, the largest lake within the Arctic Circle. Because of its large thermal mass and also because of the local topography, this lake is the focal point of a high latitude oasis, and is completely covered with ice only in some years. In warm years there is a partial thaw around late July or August, but the temperature never rises above 3°C. The aquatic plants that can withstand this harsh treatment include a few cosmopolitan mosses, which in arctic regions take to the water and become miniature versions of the grass beds of more southerly latitudes. With mare's-tail and a host of green and blue-green algae, these mosses make up the littoral zone of this freshwater Arctic lake. The herbivores include copepods and other less well-known zooplankton that are preyed upon by the opossum shrimp *Mysis oculata relicta* (Order Mysidacea) and amphipods. In some years a few anadromous fish (fish that return to the ocean after breeding in fresh water) may be found feeding on the plentiful mysids, but populations are hard to maintain when the way to the ocean may be blocked by ice for periods spanning several years at a time.

Now let us analyze the possibility of an ice-covered Martian pond. Because of increasing temperatures, the stability of ice at the Martian surface decreases as one travels from the poles toward the equator. Near the Martian tropics, the ice cover of such a pond would constantly lose water to the Martian atmosphere through sublimation. In the Martian tropics, where enough sunlight falls to keep at least part of such a freshwater pond liquid, ice is only stable deep beneath the surface, where no light can penetrate, and such a pond would have sublimated long ago.

Other lakes in both the Arctic and Antarctic are saline or hypersaline; those of the Antarctic have been studied in great detail because they contain autotrophic communities based on green and blue-green algae. These ecosystems thrive under permanent ice, where they receive hundreds of times less light than what is available at the surface of the ice. The lakes that contain a substantial salt load are meromictic—they do not turn over and there is no mixing of the surface and bottom waters. Like in the cups of the megaphytes we encountered in the tropical-alpine environment, where pectin completely inhibits convection, salt of one form or another produces the same result in these lakes. The mechanism is different, however. With pectin, the viscosity of the solution increases to such an extent that the buoyant force produced by uneven heating cannot overcome the viscous drag of the solution. In the salt lakes, the increasingly higher salt concentration as we move deeper in these lakes increases the density of the lake water and prevents the warmer deeper layers from becoming less dense than the colder upper layers. There is only a limited exchange of nutrients and oxygen between the deep and shallow waters. This exchange happens through the agency of diffusion and whatever materials are dragged up from the depths by rising gas bubbles. In these lakes the amount of light is highest under the ice and diminishes exponentially downward. Nutrients, on the other hand, are highest near the bottom, where most dead organic matter ultimately rests. At some in-between point there are both sufficient nutrients and sufficient light to sustain life. Some of these lakes, like Lake Vanda, can reach temperatures of 25°C in their hypolimnion (the deepest layer of water). This temperature probably stays fairly constant throughout the year since in ice-covered lakes such as Lake Vanda heat is lost mainly through the ice-cover, and this simply produces a thicker sheet of ice overlaying 0°C water.

Taking a cue from naturally salt-stratified ponds, engineers have designed artificial solar ponds that are optimized for the creation of a thermal gradient between the bottom waters and the top. Such a pond may be used as a source of heat, which may be withdrawn with heat exchangers to either heat a building or turn a turbine to generate electricity. A turbine system based on a solar pond must be special, since the temperature gradient is small and the usual working fluids used in turbine systems have boiling points that are too high for such a small gradient. The idea is to trap as much incoming solar radiation as possible at the bottom layer of the pond, thus heating the bottom water as much as possible. The engineering challenge is that, as the bottom water increases in temperature it decreases in density, and as in natural ponds, when this bottom water becomes less dense than the water overlying it, the pond will "turn over." In natural ponds spring turnover replenishes much needed nutrients and leads to a flurry of biological activity; in artificial solar ponds, it destroys the energy-yielding gradient and must be prevented. If the pond water becomes denser with depth, and this density gradient can be maintained even if the deeper layers are heated by the sun, then the pond will be stable against turnover.

Unlike in the Antarctic, where the dry-valley lakes have been extensively studied, in the Arctic, the few saline or hypersaline lakes that have been found have received meager attention. As of this writing, I know of only two such Arctic lakes. One is on Little Cornwallis Island, and the other is on Cornwallis Island a few kilometers to the east. The biology of these lakes is unknown.

Clearly the chemical stratification imposed on a lake by a high salt load improves the thermal environment of the lake from a biological point of view. However, the osmotic stress coupled with the toxic effects of some of the ions found in high concentrations can counterbalance this benefit; therefore, few organisms grow in these warm waters, at least on Earth. Most salt lakes have endorheic drainage, and no water flows out of the lake to eventually reach the sea. The salts that accumulate in such lakes are weathered from the surrounding rock, so that every salt lake has a unique salt composition. Some have sodium chloride as the major component and are similar to, but more concentrated versions of, the ocean. These lakes support communities that are similar to those that might be found in shallow coastal lagoons.

Although it is difficult to conceive of a salt pond that would still be liquid at the temperature of the Martian regolith (say a few centimeters below the surface), if this salt pond received even a fraction of the surface radiation it might achieve a high enough temperature (say -20 to -30°C) to retain a liquid core. There is an important complication with this scenario, however. As we saw when we discussed the possibility of freshwater ponds, the stability of ice at the Martian surface decreases as one travels from the poles toward the equator. The existence of a hypersaline pond is more plausible than that of a freshwater pond, but it is still unlikely. As with freshwater ponds, any hypersaline ponds that existed in the Martian tropics sublimated into dry salt deposits long ago. Even though such salt deposits may not contain living organisms, they are prime candidates to contain fossil traces of life.

If microbial life evolved on Mars, it must have first appeared in the lakes and ponds that were at the termini of the great Martian outflow channels, when the planet was young. Some of these bodies of water may have been relatively deep, and living organisms may have been able to evolve rapidly enough to survive the subsequent freezing of their former water world. At high latitudes (say poleward of 47° latitude), ground ice—remnants of these lakes and ponds—may be in near-direct contact with the atmosphere in a few protected spots. These spots may be canyons, the cracks of patterned ground, or maybe even the summit craters of relic pingos (regolith-covered ice mounds). These features need only be 100 or so meters deep in order to reach the ice (based on current models of the surface porosity of the planet). Such ice may be peppered with salt-encrusted microbial mats captured as if by amber. A thin liquid film on the surface of such ice-encased mats might only thaw when enough sunlight percolates through the ice at certain times of the year. Such communities would depend on light energy for both warmth and for generation of chemical energy. In effect they would be their own self-sufficient biospheres embedded in a cosmos of ice.

In this scenario, the liquid film is the critical factor that would allow life to succeed. The high concentration of sulfate salts at the surface of the Martian regolith suggests that brines that may form the liquid film are more likely to be composed of saturated sodium chloride (eutectic freezing point -20.5°C) than of saturated calcium chloride (eutectic freezing point -48°C). Clearly the lower the melting point of the surface film the longer the period of possible biological activity. On Earth salt-tolerant fungi can be shown to secrete substantial quantities of glycerol into the external medium. Such a tactic would also benefit any Martian microbes, since glycerol could prevent the surface film from freezing down to -46°C.

Deep Aquifers

Another intriguing possibility is the existence of deeply buried microbial communities similar to those that have been found on Earth. The Earth-based communities vary in their ecological makeup. However, they all require a few essentials—these include water in the liquid state, a carbon source, and a source of energy. Some of the most interesting such communities on Earth are those found 1.5 kilometers beneath the Columbia River Basalt Group—a set of flows that cover major portions of Washington, Oregon, and Idaho. In these communities the most common microbes encountered are methanogens that utilize hydrogen as the electron donor and carbon dioxide as the electron acceptor. When the groundwater contains substantial amounts of dissolved sulfates, the methane producers give way to sulfate reducers. From this description you might think that these communities are very similar to the ones we discussed when we dealt with anaerobic muds. The most striking differences are the extreme oligotrophic (low-nutrient) nature of their environment and the geochemical source of the electron donor. In some of these deep aquifers the only significant source of carbon is in the form of dissolved CO_2 and the large concentrations of H_2 may be explained as a byproduct of a reaction between the ferromagnesian silicate minerals found in unoxidized basalt and water. These communities are probably quite common when hydrogen is in plentiful supply. Even though the exact reactions that lead to the generation of this hydrogen have not been worked out, it is known that at the relatively high pH that develops when water comes into contact with fresh basalt, complexes of ferrous iron become good electron donors, and these may be the initial source of the electrons that, through some set of catalytic reactions, eventually end up in H_2. On Earth, the microbial communities that depend on this hydrogen have been touted as the first communities to be discovered that are truly independent of solar energy either for reduced compounds, carbon sources, or oxygen.

This combination of conditions may be similar to the one found at some depth in the Martian regolith, where there may be unoxidized basaltic flows in contact with liquid water or brine. If the compositon of the Martian surface is a reflection of what lies at depth, then any microbial life in the Martian aquifers would be dominated by sulfate reducers rather than methane producers.

The Radiation Environment

The incoming solar radiation on Mars is not that much less than that received in middle latitudes on Earth. In many places on Earth—from the forest floor to your kitchen's window box—plants survive on much less light than that. For instance, the sunlight received on Signy Island in the South Orkneys is 400 W/m^2, as compared to the standard value for Mars of 586 W/m^2. The seagrass *Halophila engelmanii* can grow at a depth of 90 meters in the Gulf of Mexico, where the incoming solar radiation is a small fraction of the Martian standard. Simpler plants, such as the red algae of the genus *Peyssonnelia,* may grow at ocean depths exceeding 200 meters. Sunlight on Mars does pose one problem, however—it is completely unfiltered by UV-absorbing atmospheric constituents. The levels of UV-A and UV-B radiation reaching the Martian surface cannot be matched on any of Earth's natural environments. Fortunately plants have well-proven ways to prevent UV damage or fix damage that has already occurred. These involve several types of mechanisms. The first layer of defense involves the prevention of damage by the production of screening compounds such as flavonoids. These are the pigments mostly responsible for fall color in temperate woodlands. If this level of protection is not sufficient and UV does penetrate to the sensitive leaf structures, then photoionization and photooxidation can occur. These damaging chemical reactions are mediated by several extremely reactive compounds; one of these is an energetic form of oxygen known as singlet oxygen. This energetic form of oxygen can be deactivated by a few chemical compounds, without their being caught up in a destructive chemical reaction with this rogue. One of these defending chemicals is carotene; its role is that of a bomb squad. It bumps into the singlet oxygen, disarms it by taking onto itself the excess energy, and harmlessly disposes of that energy as heat.

That is how it works most of the time, but in high-UV-radiation environments such as that of Mars, even carotene itself will eventually be destroyed by the UV radiation. Organisms can compensate for this destruction to some extent by synthesizing carotene at a faster rate, up to a point. If the level of UV damage is too high, then plants and animals must seek sites with reduced levels of UV. Plain water attenuates UV at the same rate that it attenuates Photosynthetically Active Radiation (PAR). Escape from UV in deep water or in water loaded with dissolved or suspended particulates (or for that matter in the recesses of a shallow cave) always entails at least some loss of PAR and a need to produce more efficient photosynthetic machinery. If this type of behavioral response is not enough, then the last resort is to accelerate damage-repair mechanisms. UV damage to a protein molecule normally destroys the protein's function—if not immediately (as in enzymes), then over time (as in the crystallin that forms the lens in the eyes of vertebrates). Damage to enzymes can be remedied by simply producing more enzymes; damage to the structural proteins acts slowly over time and is normally not incapacitating enough to prevent an animal or plant from reproducing. Although, if the damage is to nonregenerable proteins such as crystallin, it might lower the reproductive success of the organism.

The real damage done by UV is centered on the DNA molecule that serves as a template for all the proteins that perform the day-to-day chores inside cells. The most common form of UV damage is thymine dimer formation. Thymine is one of the four chemical compounds that go into the formation of DNA. When the DNA double helix contains two thymine molecules in close proximity, and on the same strand, UV light can fuse these two thymine molecules together forming a dimer. The enzyme that normally replicates DNA cannot read the strand beyond the damaged point, and DNA replication is thus blocked by UV damage. During the early evolution of life on Earth this must have been a constant problem, since the levels of UV-B (280-315 nm) light reaching the surface then were thirty or more times what they are now. During these early days of the ten-minute lethal tan, organisms evolved several different repair mechanisms. One of these is based on the enzyme UV-endonuclease. Its job is to search for thymine dimers in the DNA strands and cut out the affected portion of the strand while a group of three other enzymes finishes the repair process. At high UV intensities, some dimers will escape the vigilance of the UV-endonuclease and will prevent the replication of some strands of DNA. The missing proteins will throw the cell's machinery out of kilter. If the site where the damage occurred codes for a protein involved in development or reproduction, that plant or animal and its progeny are doomed. In vertebrate animals there is also the issue of uncontrolled growth of affected cells—cancer. Field experiments show that some organisms can acclimate to increased levels of UV radiation, although the level of protection these organisms can maintain does not quite keep pace with the increasing UV levels.

The Supply of Essential Gases

Besides having plenty of PAR, the Martian environment has one other advantage for autotrophic organisms such as plants; there is plenty of CO_2. Volume by volume, Mars has twenty or more times (depending on the temperature and pressure, both of which are highly variable on Mars) the carbon dioxide available to plants on Earth. On the other hand, Mars has about 1/7000 of the amount of oxygen available in Earth's atmosphere. Because of the low temperatures, water vapor is also in very short supply, even though the atmosphere is often nearly saturated with it.

You can think of the act of photosynthesis as a means of capturing and storing light energy. The energy is captured by splitting water, leaving behind oxygen, and storing the hydrogen atoms in carbon compounds (sugars). But it is misleading to think that the energy that is captured is stored only in the sugars formed during photosynthesis. Just as in stretching a spring, it is the act of separation that stores energy. On Earth, one of the products of photosynthesis—oxygen—can simply be released into a gigantic storage known as Earth's atmosphere. This is not the case on Mars. A Martian photoautotroph (an organism that uses light for energy and inorganic carbon as a carbon source) engaged in oxygenic photosynthesis would be well served by a private oxygen-storage mechanism. In this way it could utilize the oxygen it produced to derive a larger energy yield from the food it just stored; it would not have to engage in wasteful fermentation.

Our Martian photoautotroph cannot rely on a large external reservoir of O_2 in the atmosphere (like the one in Earth's atmosphere), and it must instead get some of its own oxygen to all respiring parts of its body. Of course, adaptations such as aerenchyma would come in very handy, but another possibility is a blood vascular system with an oxygen-carrying pigment. There is no reason to believe that such an adaptation can only evolve in animals; in fact some terrestrial plants, including legumes, actually produce small quantities of hemoglobins. In this case the hemoglobin is used by the plant to make its root nodules a more hospitable site for its nitrogen-fixing symbionts. It has even been proposed that the ancestral role of plant hemoglobins was that of an O_2 carrier to the rapidly metabolizing root tip.

Another possible source of oxygen for a Martian organism is the Martian soil. Part of the Labeled Release experiment showed that the soil is so highly oxidizing that it readily releases O_2 when it is wetted by water. For a native Martian organism, lack of oxygen at night would not be a problem, since the organism would be frozen soon after sunset.

Another storage problem involves the internal stores of water, which will be depleted any time there is increased gas exchange with the atmosphere. The high atmospheric concentration of CO_2 alluded to earlier may allow this gas to penetrate the cuticle of a plant in enough quantity to match the incoming PAR, without the need for the plant to keep stomata or similar structures open for very long. Therefore, most of the day the stomata can remain closed.

The Daily Freeze

The Martian year is 668.6 Martian days long, and the Martian day is 88,775 seconds or 1.03 times the length of a day on Earth. These numbers show that Mars has a slightly longer photosynthetically active period each day and many more potentially active days during the growing season—if temperature were to permit. Even at the equator, Mars undergoes marked seasonal changes due to the condensation of about one third of the atmosphere at the poles. This decrease of atmospheric pressure is a planet-wide phenomenon and affects the daily fluctuations in temperature everywhere. There would be a relatively short growing season even at the equator. The Viking lander's data clearly show that the surface temperatures maintain their "summer" levels for 100 to 150 days; after that period they drop to extremely low levels. The daily fluctuation in temperature is such that any water-based life will freeze sometime during the night unless it has a large heat capacity and a small thermal conductivity. We have discussed the cold-hardiness strategies used by insects in polar and temperate areas of the Earth and have seen that even some of the best-protected insects die when the temperature drops below -87°C. (Though a few special insects become absolutely frost tolerant in winter and can withstand liquid nitrogen temperatures.) Colder temperatures may be experienced even in summer on the warmest Martian oases, and variations of 50°C or more are the norm. The fluctuations are more pronounced within 10 cm of the ground, and since this is the region that will experience the highest temperatures during the day according to the computer models (maybe up to 10°C at 14:00 or 15:00 hours at the warmest sites), this is where the biological activity would be favored.

The leaf beetle, *Phylodecta laticollis* can withstand freezing temperatures of -30°C in the early spring, even though in the Norwegian woods where it lives, spring frosts seldom dip below -15°C. Since it relies on nucleating agents in its hemolymph to protect it from freezing damage, its response to the mild freezes it might encounter in its native habitat is to retain its winter freeze-tolerance through the spring. This example shows that activity and freeze tolerance are compatible, even though they are seldom seen together on Earth. On Earth there is always a more favorable season just around the corner, when an organism need not drag the shackles of cold tolerance around while trying to compete for mates and produce offspring. An organism on Mars would have to retain through the growing season a level of freezing tolerance that on Earth is observed only in the most cold-hardy organisms during their winter rest. Given that this freezing tolerance is based on high levels of polyols and the presence of ice-nucleating proteins and antifreeze proteins, theoretically, these organisms could retain their freeze tolerance through the growing season at the expense of some reproductive potential. In this regard it is appropriate to remember the plight of the chironomid *Polypedilum vanderplankii* in the African temporary rock pools. This insect is given little warning before the pools dry up completely, and yet, in its state of almost complete dehydration it can withstand immersion in liquid nitrogen or brief exposure to 104°C! During its larval life this insect must always be prepared for the unpredictable bouts of dryness. Once the pool fills with the next rain, it rehydrates in a couple of hours and goes about its business oblivious to the intervening spell of near death.

The Yearly Freeze

The winter-cold-tolerance strategies seen in terrestrial plants and animals that survive the coldest winters of the North American and Asian continents would also work on Mars given three provisos. First, the degree-day budget for reproduction of the organism (from the break of winter dormancy to the ripening of propagules) must be achievable at least once during the life of the organism. We have seen that in the Earth's arctic regions the plants are—with but a few exceptions—all perennial. The reason for this is the cost, in terms of degree-days, of producing new leaves and propagules, when the whole degree-day budget could be used to produce propagules of some kind. Second, there must be enough degree-days every year to allow for winter hardening of the organism (this is a more stringent requirement than the first). For instance, on Earth the currant *Ribes nigrum* can properly harden only after 5 months of growth; shorter periods lead to increased mortality until, if the growing season is less than 2 months long, all plants die in winter. The third proviso is due to the desiccating force of the sun, even in winter. On Earth, in a region where drought is a concern, the environment dictates that the plants relocate the valuable resources they have in their leaves to underground storage organs, lest they lose these resources to the four winds during the inhospitable seasons. Such an adaptation applies to some temperate deserts on Earth as well as to Mars. Consequently, geophytes are common in some of the most seasonally inhospitable places on Earth. They are not common in the polar regions because of the difficult task of penetrating frozen ground. In fact, the two quintessential bulb families—the lilies and the amaryllis—do not have any members in the High Arctic. Any Martian plant would have to spend the winter in a protected place, preferably underground, or have a thick impervious armor, like that of *Dioscorea elephantipes* a wild yam of the southern African semideserts. Such an armor might help it survive the onslaught of UV radiation in a season when all its repair mechanisms would be frozen and unable to do their job. The lack of permafrost immediately below the surface is an advantage in terms of winter survival in a land without snow. A Martian plant in a sandy area could have underground storage structures without the difficulty of penetrating permafrost. Luckily, sand seas seem to abound on Mars.

The Low Pressure

Low pressure is an aspect of the Martian environment that is completely foreign to Earth-based life. We can get an inkling of the effects of low pressure by studying the biology of high mountains, as we have done in chapter 10. A point emphasized there and here also is that sunlight-induced desiccation is a particularly severe threat as the atmospheric pressure decreases. This effect is due to the decreased boiling point of fluids as pressure decreases. When we talk about water on Mars it must be understood that this water is most likely in the form of ice or deeply trapped brines, since pure liquid water is unstable under such low surface pressure. At 10 mb (that is a high pressure ridge in the Martian atmosphere), an open container of pure water will boil at 7°C. Of course, any present or future Martian organism is not an open container of water; a closer analogy would be a very tightly sealed rigid bottle. The water in such a bottle will not boil, even above that temperature, until the pressure increase inside ruptures the container. However, many plants on Earth can stand internal pressures of several atmospheres, and it would not be hard for a Martian organism to do the same. In this regard it is useful to remember that many plants have an active water-uptake-and-excretion mechanism in their root systems whereby the root cells secrete ions into the xylem. The increased ionic concentration in the xylem draws out water from the root cells and this water exerts a pressure of up to 5 atmospheres. When the pressure becomes high enough, water is released through special openings in the leaves called hydathodes. Plants from grasses to irises to strawberries have this ability, and most people have observed the little droplets of water at the tips of the leaves in the early morning, but have mistakenly dismissed them as dew.

One other possible effect of the low pressure is worth mentioning. There has been an intriguing suggestion that decreasing pressure will increase the cold hardiness of some plants. The evidence to back this assertion is scant. But the possibility exists that, since the membrane phase transitions play a critical role in producing freezing damage, suppressing these phase transitions will help plants survive freezing temperatures. As the temperature drops, cell membranes undergo a transition from a liquid crystal state to a more gel-like state. This transition is accompanied by an increase in the density (or a decrease in the volume if you prefer) of the membranes. It is possible that, by relieving the pressure around the membranes, the less dense phase is favored, even at lower temperatures. Thus low pressure may actually enhance survival at low temperatures, at least for certain plants.

The Lack of Water

Clearly Mars is a desert by any standard. The soil contains less than 1% moisture by weight, and it never rains. Nevertheless, the Viking landers photographed summer frost accumulating on the shady side of rocks and there is a clear daily depositional cycle; although if liquid water is present, it is very fleeting. One possible source of water is the relatively moist Martian atmosphere, and as we have seen, on Earth there are some organisms that have behaviors, mechanisms, and structures to take advantage of atmospheric moisture. With a much reduced pressure, the amount of ventilation necessary to pass enough Martian air through a water-extracting structure would be prohibitive. Any such structure would have to be external to the organism, so that the ventilating would be done by the gentle breezes near the ground. Since such a mechanism would have to capture the moisture while the relative humidity was sufficiently high, it would have to operate autonomously during the cooler part of the day. Likely periods would be just after sunset or just before sunrise, when the organism would most likely be incapacitated, if not actually frozen. The highly evolved and efficient mechanisms of the ambionid beetles, the firebrat, and the mallophagan lice (discussed in chapter 3) would fail under such circumstances, because they are internal. The less evolved mechanism of the desert cockroach (*Arenivaga investigata*) might be a more useful example of a mechanism that might work in the Martian environment. This roach uses eversible and inflatable sacs in its mouth to collect atmospheric moisture; these sacs are covered with hygroscopic hairs on which the moisture condenses. When the hairs are loaded down with condensation, special glands release a liquid that forces the hairs to give up their load of water. A similar mechanism can be postulated for a Martian organism. The hairs that adsorb the water vapor could do so as night fell and the animal became dormant. At daybreak, the frost that had accumulated on the hairs overnight would be turned back into vapor. The hairs would then adsorb this vapor and, when the animal could move, the mouth would close, preventing the loss of the vapor back to the drying air. Other mechanisms that may be effective in the Martian environment include having special hairs on the legs that act as capillaries, drawing up water toward the animal's mouth (as we have seen in the terrestrial crabs). Another germane example is provided by wolf spiders, which can abstract soil water held tightly by capillary forces. Such a mechanism would have several advantages for a Martian organism. First, it does not require any immediate expenditure of energy on the organism's part, and secondly, it would be relatively rapid. Speed is essential to take advantage of the fleeting presence of liquid water (if it is present at all). The organism would have to move from its feeding site to a site of maximum frost deposition before it loses its motility. Once there, the special capillaries could absorb water as the frost melted in the early morning.

Another example from Earth provides an even better model for a Martian frost-capture system. The web of the European garden spider has glue droplets on the spiral capture threads. These glue droplets are not only very effective in capturing prey, but also absorb water vapor from the air. These spiders maximize their water intake by eating the webs during the most humid part of the night. Such a water harvester can be envisioned living on Mars. An organism with such a capability could set its water trap on the west side of a rock the previous afternoon and then move to spend the night on the east side. The following day, the Martian spider would thaw out and move about while leaving its water trap safely protected from the drying sun by the rock's shadow. The Martian spider would then, from the warmth of the sunlit side, reel in its net, or make short trips into the cold shade to harvest the water-swollen droplets of glue. Whether such a mechanism would work on Mars is open to debate. There the atmosphere contains significant water, but the condensation point is normally lower than -70°C. The water vapor content of the air in the Northern Hemisphere of Mars during spring and summer can be relatively high, and coupled with the relatively high pressures (around 10 mb), it may provide enough water for such a water-collecting approach.

Martian plants could harvest water from the permafrost that is proposed to exist over much of the planet. The problem for organisms is that the depth to permafrost is predicted to be inversely proportional to the warmth of the surface. In the warmest areas of the planet, permafrost is predicted at a depth of several hundred meters. On Earth, only a few exceptional trees may send roots that deep, and in these cases the trees had sporadic rainfall to supply them with water while they tried to reach the deep water tables. On Mars the problem is much more complicated, because the soil becomes too cold to allow metabolic processes to proceed at any but the slowest rates. Any Martian plant hoping to tap the hidden permafrost would have to produce its own cellular heat to spur root growth. Once its roots reached the level of the permafrost, the metabolic heat might be used to melt the ice and allow the plant to absorb liquid water. Like on Earth, the permafrost level will not be a flat plane, but will be as topographically varied as the surface, with hills where large surface rocks lie, and deep valleys where the surface is loose sand. A Martian plant would then grow the same way long-lived perennials grow on terrestrial deserts. They would start life in the sandy areas supplied with moisture by the internal dew of the sand, and would then grow laterally toward the permafrost hills underneath large rocks.

Alternatively, Martian plants could take on a rock-base habit, as we have seen in the aeolian zone of high mountains on Earth, or they may grow in the shape of a rock as do the wild yams (*Dioscorea* spp.) of the South African deserts. If a Martian plant grew in the form of a rock with a corky rough surface it would be well poised to absorb the water deposited by the night's frost. Like the lobelias of the giant African volcanoes, it would have aerial roots penetrating the absorptive material at the plant's surface or trichomes in the furrows of the bark, ready to catch any frost that would melt and turn momentarily to water before evaporating. Taking a cue from the hellebores and the snowdrops, such a plant might have a top rosette of large leaves that would flop down as the cold became severe. These leaves would then serve as a lid, trapping the frost deposited on the corky bark, and the next morning giving the plant enough time to absorb the water before it evaporates.

Martian animals could also have evolved to mine deep brines, just as some ants and termites mine deep, brine-soaked sands in the deserts of Africa in order to keep their nests at a high relative humidity. Termites have been observed to excavate vertical tunnels to reach water at depths of 40 to 60 meters.

Animals in this harsh vision of life would have to seek out plants not only for food, but also for oxygen. On Earth this behavior is common in aquatic environments, and just as the larvae of a European mosquito, *Mansonia richerdii*, seek out the roots of cattails in order to pierce the roots' epidermis and derive oxygenated air from the aerenchyma, our putative Martian animal could derive oxygen from the roots of a Martian plant. On Earth, several different sorts of insects accomplish this feat with the aid of piercing spiracles. It thus is a trait that evolves readily under the right circumstances. Another option for Martian animals would be to have photosynthesizing algae in their own cells, like the zooxanthellae in the cells of corals. On Earth, several kinds of invertebrates have such intracellular symbionts. Sea slugs with such algae can be seen trailing bubbles of oxygen as they move. However, the role of these intracellular symbionts is probably not to provide oxygen, but instead, to provide camouflage and nutrition. In the case of corals, these symbionts even act as solar-powered chemical pumps that, by removing dissolved CO_2, help the coral hydroids deposit calcium carbonate.

The Martian Regolith

The current concept of the Martian soils paints them as highly oxidizing mineral soils with a composition similar to that of weathered basalt sands, as might be found on Surtsey or in Hawaii. The other contender is a soil derived from basalt, but weathered by water millions of years ago and having a smectite clay similar to montmorillonite. We have seen that plants have difficulty growing on montmorillonite hills, but as far as water availability is concerned, montmorillonite is much more likely to support plant life than unaltered basalt. At a plausible partial pressure of water vapor in the Martian atmosphere this clay will hold 20 to 120 times the amount of water that basalt holds at the same temperature and pressure. The water bound to soil is adsorbed to the internal and external surfaces of the soil particles. This ultrathin coating (often just a few water molecules thick) is held very tightly, and (as we saw in chapter 6) the force with which it is held is called "the matric potential." It is a potential because energy must be spent and work must be done to remove this water from the matrix. In this sense, it is no different from the water in a concentrated salt solution. The one clear difference is that in a salt solution there is a lot of water. In such a solution, water surrounds the organism or its uptake structures, whereas in soil, the water film is discontinuous, and the discontinuity can bring water uptake to a halt even when the matric potential itself would suggest that water is still available. This predicament affects mainly plants, because of their reliance on the cohesion-adhesion-tension method of water intake, which requires a continuous daisy chain of water molecules.

On Earth, some arthropods derive a substantial fraction of their moisture intake from the substrate through which they burrow. This is the case with the larva of the tiger beetle *Cicindela marutha*. It lives in a vertical burrow that may be 30 cm deep. The larva's head and thorax form a plug at the top of the tube to prevent the entry of any predators. This larva has on its dorsal surface hooks that help it wedge itself in its burrow. Both the larvae and the adults have powerful mandibles and are predaceous. The larva waits for its prey to bump into its open jaws and then tries to drag the victim into the burrow. Tiger beetles prefer sandy areas both as juveniles and as adults. The larvae derive a significant fraction of their moisture intake from eating the sand of the walls of the burrow. Of course, the amount of water obtained depends on the efficiency with which the digestive tract can extract water from the sand. As we have seen with those organisms that have the internal machinery for the extraction of water vapor from the air, once the water is in the digestive tract, it is much easier to incorporate it into the tissues. For one thing, the desert-arthropod cuticle can be extremely impermeable because of an external layer of lipids and hydrocarbons, and once the body's entrances and exits are sealed, it becomes the perfect container.

For a Martian animal living in areas without surface frosts, mining for water bound to clays or found as deep frost in sandy areas would be the most likely adaptation to the lack of water at the surface. But there are several thorny problems with this scenario. The first problem is that of surface abrasion of the impermeable cuticle which is a necessity in such an environment. Such abrasion destroys the impermeability of the cuticle, and would either require periodic molting, or reapplication of the surface hydrocarbon layer, or both. To be able to penetrate deeply (30 cm or more) into the loose Martian sands would require the ability to operate at very low temperatures. Metabolic heat generation would come in handy in this instance also, but again, the availability of oxygen will limit the rate of any heat production. In theory, the Martian soil itself could provide a way out of this predicament. From the Labeled Release experiment we know that at least the first few centimeters of Martian soil are so reactive that when the soil comes into contact with water, O_2 is released. Therefore a Martian animal could trade one precious commodity (water) for another (oxygen). The question then becomes whether the amount of released oxygen would justify the amount of water spent. Such a mechanism would have to be based on a very efficient reuptake of the secreted water. Even then, the amount of releasable oxygen may be so low that it will not provide a net benefit to the organism.

As we have seen, it is likely that the Martian substrate is highly oxidizing. An organism could derive almost all of the energy available through oxidative metabolism by eating the soil! On Earth it has been shown that several microorganisms, including bacteria and fungi, can use oxidized iron compounds as the terminal electron acceptors. Sometimes the iron is reduced by some intermediate, and the microbe does not directly benefit from the extra energy released by the reaction; but it is thought that some other microbes can directly utilize the oxidized iron compounds in energy production. For the most part, oxidized iron found in soils is in extremely insoluble forms, and how these organisms can include such insoluble compounds in their metabolic pathways is somewhat of a mystery.

A Martian animal could therefore derive not only water but also an oxygen-like electron acceptor from eating the soil, and such an organism would be highly adapted for a fossorial life (like the moles and mole crickets of Earth). It would have to be relatively large—to have a significant thermal mass—and to be well covered with hair for insulation and protection from abrasion. It would have digging claws and a hard cuticle impervious to water.

Humans and Life on Mars

For humans to exist on the Martian surface, a reasonable Martian agriculture would have to be established. I believe that reasonableness would be the key to the success of such an endeavor. It is unrealistic to think that either multinational companies or industrialized nations would treat a human presence on Mars the same way they treat the find of a large oil field. There is no commercial incentive to the establishment of a Martian outpost, and there won't be for a long time to come. At least for the foreseeable future, the only reason to establish an outpost on Mars would be for scientific purposes. In some circles this is no reason at all and is just another instance of wasteful governmental spending. I must assume that anyone who has read this far is not of that persuasion.

The men and women who will eventually live and work on the red planet will undoubtedly have to bring with them their own crops and animals to use in this new land. Unlike many previous journeys of exploration and settlement undertaken by peoples around the world, few of the plants and animals we now use for food, fiber, and construction would be useful on the Martian surface. I am not dismissing our ability to produce Earth-like habitations, self-reliant in most respects. But in recent years, attempts to create such environments have shown that biological systems do not behave like gears in a clock; they are variable, and any design must take this variability into account. Add to this variability the one imparted by the Martian environment itself and the ever-present chance of breakdown of the mechanical and electrical components of an artificial habitat, and you have a daunting design task that requires several levels of safeguards and backup systems. It would be best if the plants and animals chosen to accompany humans on this journey are those that are most resilient and robust, and can survive conditions on Earth approximating those of the undiluted Martian environment. Both plants and animals, even those grown in a greenhouse environment, will have to be insensitive to some level of nightly frost. A relative insensitivity to nighttime oxygen deprivation would be an advantage, although the cold temperatures on Mars will minimize the effects of low oxygen. Be that as it may, aerenchyma would be the way to go for any artificially bred Martian plant of terrestrial parents. The cost in terms of labor and materials for a greenhouse to sustain a full complement of people in a Martian station would dictate the use of fast-growing plants and animals. In this way, the structure can be as small as possible and yet produce the necessary food and other commodities. Drought tolerance is not as high in the order of priorities as the previously mentioned requirements, but some drought tolerance must be included in the design, a failure of the water distribution system could spell disaster in such a self-contained system.

The high levels of iron and other transition elements might pose a problem once the mineral soil is amended with organic matter and watered. As we have seen, wetland plants must have mechanisms that prevent metal ions from entering their root tissues. Many of these plants have aerenchyma that allows oxygen to escape and precipitate the iron just outside the roots. There is a good correlation between the extent of aerenchyma in root tissues and the ability of wetland plants to survive elevated concentrations of ferrous iron in anaerobic soils on Earth. Since most wetland plants on Earth rely mainly on atmospheric O_2 to produce this effect, the efficacy of aerenchyma as a preventive measure against iron toxicity is an open question for Martian agriculture. If the load of reduced iron is too great, even adapted plants will eventually have to produce new root tissue as the old tissue becomes clogged with precipitated oxides of iron. In this regard, conditions might be similar to those in some wetland soils we have discussed.

Another unanswered question deals with the level of salinity in the Martian soil. If salt does pose a problem, then another selection factor in choosing the initial candidates will be a certain amount of salt tolerance. A related problem is that of high levels of hydrogen sulfide that may be produced by sulfate-reducing bacteria from the high levels of sulfate salts that seem to be present on the Martian surface. If there is enough water, the soil can be washed; otherwise, the whole agriculture system can be based on a frugal hydroponic system. Water will be at a premium, and the flow of the nutrient solution will have to be carefully controlled. Low-volume misting systems have been proposed, and they may work if all the pumps, sprayers, and controls work appropriately. Just as with large-scale fuel cells used for electrical energy production, it is a great engineering challenge to maintain a system based on fluid flow working when the environmental temperature plummets.

In every other venture of human settlement, we have introduced foreign species into existing ecosystems; Mars is different. If life is absent from Mars, then all those things we take for granted until they are gone (and then we are surprised when a natural resource vanishes "overnight") must be introduced along with the plants. Many plants have associated microbes that provide them with nutrients from the soil, or that fix the nitrogen in the air into a form plants can use. Other microbes decompose the dead plant material that accumulates or destroy the germination inhibitors in seed coats. All these organisms—and many others of whose actions we are blissfully oblivious—would have to be introduced. The problem is one of approach. It is certain that clumps of soil from the High Arctic, the aeolian zone, and Antarctica probably will contain most of the material recycling functions and organisms required. These organisms are the ones that have a fighting chance of surviving in the new environment. This approach does not provide us with the certainty that no pathogenic organism would be introduced. The only way to make sure of that is to take inocula from pure cultures, and with this approach, the chances of overlooking a critical component of the system we are trying to create are great. Consequently, several possible crops should be chosen on Earth, and every reasonable effort should be made to grow them under conditions approximating the ones to be encountered in a Martian greenhouse. A Martian greenhouse would have to undergo an initial bout of succession, just like any newly exposed mineral soil on Earth. The first set of crops must be able to grow without organic amendments—they will become the organic amendments for future crops.

There are several adaptations that will benefit any plant trying to grow and reproduce in the Martian environment. The more organically rich the soil in a Martian greenhouse, the larger the oscillations of oxygen concentration will be. Control over such a system will be coarse at best, as is the case with most multispecies biological systems. The coarsest level of control will be the amount of organic matter introduced into the soil—the organic matter will be the main source of food for the respiring soil microorganisms (the main consumers of oxygen). The second control is the amount of water used in irrigating the plants. This control is finer than the previous one, but it is not linear; the highest respiration will be found at intermediate levels of irrigation. Since water will be scarce (at least initially), the irrigation will be at frugal levels, and the plants that are chosen must be able to withstand a certain amount of wilting stress. They should also be able to withstand a certain amount of oxygen deprivation as the sun sets and before the temperature drops to a low enough level that cellular respiration ceases. A relatively high level of resistance to oxygen deprivation would also allow such crops to survive prolonged periods of lowered light intensity, as was observed during the great Martian dust storm of 1973.

A few plants have been shown to withstand complete anaerobiosis for at least several days. Some of these plants grow in areas where freezes during the growing season are common. They belong to the following families: quillwort, clubmoss, saxifrage, buttercup, mint, mustard, rose, iris, lily, amaryllis, grass, sedge, and rush. Most of these families have members that are currently used as sources of food or fiber, and the rest are used as graze for domestic stock. I am sure there are other plant families that have members that meet the requirements, but these have yet to be tested.

The matter of animals as a source of protein in such an environment is more problematical. No mammal or bird is likely to meet the anoxia requirement, though many will meet the cold requirements, if well fed. Insects that could serve this role include the leaf beetle, *Phylodecta laticollis*, given its cold hardiness while in an active state. Its tolerance to anoxia is not known, but other chrysomelids (leaf beetles) live and feed on submerged parts of aquatic plants. We should not let the food biases of Western cultures hamper the search for the most appropriate species for a Martian agriculture. Insects, as well as other invertebrate sources of protein, are important in many cultures, and I have the feeling that this will be the case on Mars.

What twists and turns evolution might take on a slowly freezing planet no one can tell, but as we have seen, on Earth living things seem to adapt to even the most inhospitable conditions this planet has to offer. May this not also be the case on Mars?

12 Lessons Learned

When trudging through a salt marsh one notices the absolute dominance by a few species of plants; these plants have triumphed over the limitations (high salinity and low oxygen) imposed on them by the environment. If these plants had not adapted to this harsh environment, there would be a less prolific but possibly more diverse community in its place. If the salinity is further increased, and oxygen further decreased, then one enters the realm of the limits to life. In certain areas of the high marsh the salinity is so high that salt encrusts on the soil surface, and since there is no flushing by the tides, oxygen does not penetrate the black mud. Such areas have been called salt flats or salt pans; in them the inhibitory effect of the combined stresses is much greater than that of either stress alone. This is the realm of the glassworts (*Salicornia* spp.). Here, if the glassworts were removed—-or had never evolved in the first place—the area would be completely barren. (In that case there would still be microbes, but even they have their limits).

Simple or Complex, Which is More Robust?

When we deal with extremes, we must differentiate between the types of organisms involved. For instance, the limits at which microorganisms cease to function are much broader than those for any other form of life. In these pages we have encountered microorganisms in permanently flooded soils, on the surfaces of barren rocks, and even in saturated brine. More complex organisms approach these limits, but they never surmount them. Even though we may speak of bacteria, fungi, and algae as simple organisms, it must be recognized that their biochemical repertoire is extremely varied, and in many ways, surpasses that of more complex life.

When we turn to multicellular organisms, we must realize that the limits encountered depend greatly on the evolutionary path already taken by a particular group. As we have seen, if a plant has a woody habit, then cold can be one of the most severe limiting factors, simply because of the degree-day cost of making wood.

Each vascular plant type has its own limit. For instance, much has been written about the general concept of the treeline; that is, the boundary between forests and either low-growing vegetation of some kind or barren surfaces. We have already discussed the north temperate treeline, but there are many other more subtle "treelines." A case in point is the temperate limit of growth of mangroves that we discussed in chapter 5.

Available Options

There are many solutions to the problems faced by living things; the solutions adopted by organisms need not be the best possible, they only need to be good enough to allow the organism to compete for resources. (You don't need to win every hand, just enough to stay in the game.) As environmental conditions become more extreme, organisms become fewer and more scattered. Therefore, at least for sedentary organisms, biological competition for resources becomes a less important factor in determining population numbers than the factor of physical stresses. In extreme environments, fewer and fewer solutions cut the mustard, and even these few solutions are barely adequate.

In most of the extreme environments the trend toward fewer and fewer solutions can be explained by several factors. One of these factors is the newness of some of the really extreme habitats; for example, High Arctic areas that were exposed by glacial retreat and have been so for a relatively short time. This factor gives organisms little time to colonize suitable areas, especially in places like the High Arctic where reproduction may be a prolonged multi-year affair. We find time and again that on newly exposed land such as the island of Surtsey discussed in chapter 4, the rate of colonization by organisms is set by the availability of propagules, the order in which they arrive, and last but not least, a favorable environment.

We will now delve into the types of solutions most often encountered in extreme environments. First we will deal with the manipulation of size, which will be shown to be a way of avoiding the extremes of the environment. Next we will deal with the chemical limits to life. Here avoidance and tolerance both play a role.

Size

On Earth, almost all offspring start out external life smaller than their mother (there are a few exceptions, such as some of the viviparous flies), and juveniles have a definite size disadvantage in certain types of extreme environments. In environments where daily temperature extremes are the norm, a large thermal buffer due to a large body mass is a great advantage, since it buffers the temperature fluctuations. This adaptation helps the megaphytes of the high Andes and African volcanoes escape freezing temperatures. (They also have other adaptations, however.)

In environments where water is scarce and precipitation infrequent, large size means a larger water store, and a greater chance of surviving until the next rain. I once collected a pad of *Opuntia ficus-indica* that had washed up on the beach (who knows how long it had been bobbing along), and I kept it in a container protected from rain but exposed to the scorching afternoon sun for almost a year. Once I placed it on the surface of the ground, it produced a new pad within a week. No wonder the opuntias are such successful cacti! This principle does not apply only to plants; for example, aestivating animals, be they lungfish or frogs, have a better chance of survival if their water stores are large.

Another extreme environment where large size provides a great advantage is the anoxic mud of swamps. Since shoots emerging from a deeply buried root system must grow up to the surface while deprived of oxygen, they must depend on the wasteful process of fermentation to produce the energy they require. If the plant exhausts its starch reserves before reaching the surface of the mud and sunlight, it will die. Plants adapted to this environment have large starch-storage organs in the form of tubers, corms, or rhizomes. This may explain the ridiculous sizes of the corms of the giant swamp taro (*Cyrtosperma chamissonis*), some of which have weighed in at 180 kg.

The examples I have cited are meant to illustrate that in many extreme environments, a mature individual may have a good chance of success, but a smaller (immature) individual runs a great risk. Some extreme environments are barren, or almost barren, because the chances of successful colonization by plant propagules are nil. In such environments, we sometimes find relic populations of old individuals (of about the same age) and no trace of juveniles.

Large size (and the concomitant reserves of heat, water, and food) allows organisms to dampen oscillations in environmental variables: oscillations that would otherwise prove fatal. Such an approach works best for sedentary organisms such as plants, or organisms that survive the harshest conditions in a torpid state. However, another way to obtain the same dampening of oscillations is by burrowing or hiding in the crevices of large rocks, using the substrate as a thermal or humidity buffer. Such an approach is taken by the highest communities found on the highest mountains. There, a mere passing cloud might presage the coming of high winds and plummeting temperatures within a few minutes; this cue sends the small jumping spiders and the springtails scurrying for cover underneath rocks were warmth lingers a little longer than in the open.

The extreme conditions of harsh environments impose physical (as opposed to biological) constraints that limit the number of evolutionary solutions available to organisms. For instance, to forage for food, or simply to develop and grow, is impossible if an organism is frozen. In the few extreme environments where organisms are active below 0°C, their internal fluids must remain, well, fluid. There are only a few approaches available to organisms intent on keeping their mobility at subfreezing temperatures. These approaches include the following: producing nucleation inhibitors, supercooling, and combining the former two approaches. In all cases, there must also be changes that will allow the cell membranes to remain in a fluid state as temperatures drop. The only other option for such organisms is the generation of internal heat; to be effective, this approach requires a steady source of food.

In other cases the explanation for the limited number of evolutionary solutions must be that these extremes truly tax the creativity of nature. In these pages I have attempted to show that what is taken as common knowledge regarding the limits to life is just not so. Maybe the most obvious, but also the most important, of these limits to life is the lack of water, since organisms have an absolute need of liquid water to serve as the internal solvent. A large part of this book is concerned with what organisms do when water is a rare commodity. Another important point to be made is that, as organisms and their interactions become more complex, they can tackle situations that would have meant total annihilation to their evolutionary ancestors. This increased performance of complex organisms depends on a higher rate of energy consumption; this high rate then becomes essential to sustain life.

Chemical Limits

Water is the solvent that allows most living processes to proceed. Even though water has one of the widest liquid temperature ranges of any substance, it still freezes at a relatively high temperature. If this fact were immutable, organisms would have no chance of surviving subzero temperatures, but organisms can modify the physical properties of their internal solvent through a variety of mechanisms. How far can organisms modify the properties of their internal solvents? We have taken a glimpse at the answer when we discussed the halophiles. In some of these organisms glycerol becomes a major component of their internal solvent, and special proteins can control when and where ice forms. To carry out their functions, cells must not only have liquid interiors, but they must also have a fluid boundary between themselves and the outside world. This boundary is, of course, the cell membrane. We have seen that most organisms opt for a lipid bilayer for this structure, but some others use hydrocarbon ethers. Such a variety of building blocks means that there is a lot of leeway in the properties of biological membranes. The liquid-crystal-to-gel transition point we spoke of in the previous chapter can be manipulated so that it doesn't occur until relatively low temperatures, for instance. In order to survive, a cell must have, besides a fluid boundary and a fluid interior, enzymatic machinery that works under extreme conditions. All enzymes have an optimum temperature; outside this optimum, their efficacy as catalysts decreases. However, some lichens can be shown to still fix CO_2 at -24°C. So, even at this low temperature some enzyme systems can still function, although at reduced rates. At high temperatures, enzymes lose the three-dimensional structure that gives them their catalytic function, and the upper temperature limit to life seems a much more concrete boundary than the lower temperature limit. Even here, however, biotechnological research has shown that isolated enzyme systems, once protected by bonding to polymeric substrates, can achieve high catalytic activities at elevated temperatures.

All free-living organisms must degrade high energy compounds into low energy compounds and use some of the liberated energy to grow. For life as we know it, high energy compounds equate to electron-rich compounds. What constitutes an electron-rich compound is dependent on the immediate environment of the organism, especially the available electron acceptors. For instance, many organisms can use acetate as an electron donor when a strong electron acceptor (such as oxygen, nitrate, or sulfate) is present. When these acceptors are absent, only one organism (a bacterium) is known to be able to oxidize acetate (using elemental sulfur as the electron acceptor) to produce ATP.

Our Lessons

The weather on Earth has been a powerful force in human history. We have learned many lessons about survival in the past few thousand years. These lessons have been woven into legends and lore as a way of passing them down to new generations—whether we listen is up to us.

The Icelandic sagas are an example that can be verified by archeological evidence. They speak of two Norse settlements in Greenland started in 985 A.D. Like all other Norse settlements, these were based on a pastoral culture. In this culture, cattle, sheep, and goats were the basis of the economy and hay was the primary crop, with a few grains such as barley and oats grown for human consumption. These staples were supplemented by fishing and by hunting—mainly migrating harp seal (*Pagophilus groenlandicus*) and caribou. The western settlement was probably destroyed by skirmishes with the Inuit who were migrating into the area by the fourteenth century, but the larger eastern settlement met a different fate.

As the weather deteriorated during the centuries following the colonization, the Norse settlers had to keep their livestock inside for longer and longer periods, and crop failures became more commonplace. Fishing became more difficult because of longer periods when the sea ice hugged the coast. The culture and the technology available to these settlers were not sufficient to cope with these changes, and the settlement died out. What sealed their fate was their inability to modify their culture and technology along the lines of the successful Inuit. During the same period, life in Iceland also became harder. Barley became more difficult to grow, until by the sixteenth century its cultivation had died out. The cultivation of barley failed because this grass requires 850 degree-days to mature when the annual precipitation is 200 mm. If the weather turns wetter, then barley requires a longer growing season, and it has been estimated that for every 100 mm above 200 mm of precipitation per year, barley requires an extra thirty degree-days. During the period of worsening climate, some farmers gave up on the cultivation of the traditional crop and switched to the collection of the seeds of the plentiful native lymegrass (*Elymus arenarius*). Such adaptations are seen in times of desperation; they are, however, quickly forgotten once the climate improves.

There are many other examples of settlements or even whole cultures that were destroyed by worsening climatic conditions. In the puna of Bolivia, the Tiwanaku culture thrived a thousand years ago by exploiting an ingenious farming method that tapped the heat-storage capacity of water. By building a maze of canals intermeshed with their raised fields, these high-altitude farmers were able to keep their crops from being damaged by severe spring frosts that limit plant life at this altitude. The trouble with this ingenious solution is that it requires large amounts of water, and in years of drought the early frosts were probably devastating. Based on climatological records, it seems that, between 950 and 1000 A.D., recurrent drought totally disrupted the way of life of the Tiwanaku and caused their insightful technology to be lost for centuries, until its recent recovery by the archeological team of Kolata and Rivera and their colleagues.

These examples should suffice to show that extreme environments deserve more than just academic interest. Throughout most of human history, calamity inflicted by nature has been a much more significant cause of human suffering than self-inflicted strife. Even now, when we have the means to do untold damage to ourselves as well as to the rest of creation, an impending ice age would create much more dislocation than most man-made events. We live on an unpredictable planet, and the lessons we learn about coping with extremes might prepare us for the future.

APPENDICES

— ⌘ —

Each species has a taxonomic code associated with it that may represent its family, order, class, or phylum. The common and scientific names corresponding to the taxonomic codes are found in the taxonomic code appendix. The choice of taxonomic level is based on the general availability of easy-to-use field guides.

— ⌘ —

Plant Species Mentioned

Latin Name	Common Name	Taxon
Acaena trifida		rosa
Acalypha dikuluwensis	three-seeded mercury	euph
Acer rubrum	red maple	acer
Acorus calamus	sweet flag	arac
Agrostis perennans	perennial bent grass	poac
Agrostis stolonifera	creeping bent grass	poac
Agrostis tenuis	bent grass	poac
Allenrolfea occidentalis	iodine bush	chen
Alopecurus alpinus	alpine foxtail	poac
Alyssum ovirense	alyssum	bras
Alyssum wulfenianum	alyssum	bras
Ambrosia dumosa	white bursage	aste
Ammophila arenaria	European beachgrass	poac
Ammophila breviligulata	American beachgrass	poac
Amorphothalus titanum		arac
Androsace chamaejasme	rock jasmine	prim
Arabidopsis thalinum	mouse-ear cress	bras
Arabis alpina	rockcress, alpine	bras
Araucaria	araucaria	arau
Araucarioxylon		arau
Arenaria patula	sandwort	cary
Argyroxiphium sandwicense	silversword	aste

Armeria maritima var. *halleri*	sea-pink	plum
Artemisia herba-alba	artemisia	aste
Artemisia skorniakowii	artemisia	aste
Atriplex confertifolia	four-winged saltbush	chen
Atriplex hastata	saltbush	chen
Atriplex nuttallii	Nuttall's atriplex	chen
Avicennia germinans	black mangrove	verv
Azolla pinnata	mosquito fern	azol
Azorella spp.	yareta	apia
Batis maritima	saltwort	bata
Becium homblei	copper flower	lami
Bulbostylis mucronata	watergrass	cype
Cakile edentula	sea rocket	bras
Calamagrostis deschampsioides	reed grass	poac
Calamagrostis tolucensis	reed grass	poac
Calceolaria spp.	calceolaria	scro
Calla palustris	wild calla	arac
Callitriche spp.	water-starwort	call
Caltha palustris		
Cardamine spp.	bitter-cress	bras
Carex lyngbyei	Lyngby's sedge	cype
Carex microchaeta	sedge	cype
Carex moorcroftii	sedge	cype
Carex pachystylis	sedge	cype
Carex stans	sedge	cype
Carex subspatheca	sedge	cype
Carex ursina	sedge	cype
Carpobrotus edulis	Hottentot fig	mese
Cassiope tetragona	arctic white heather	eric
Casuarina spp.	Australian pine	casu
Ceratiola ericoides	Florida rosemary	empr
Ceratoides compacta		chen
Ceratoides lanata	winterfat	chen
Chenopodium rubrum	goosefoot, red	chen
Chenopodium salinum	goosefoot	chen
Chionodoxa	glory-of-the-snow	lili
Chionoscilla		lili
Chloris gayana	rhodes grass	poac
Chorispora	Himalayan mustard	bras
Christolea himalayensis		bras
Cicuta virosa	cowbane	apia

Claytonia perfoliata	miner's-lettuce	port
Claytonia virginica	spring beauty	port
Cochlearia spp.	scurvy grass	bras
Colliguaja odorifera		euph
Colobanthus quitensis	pearlwort	cary
Copiapoa cinerea	cactus	cact
Copiapoa haseltoniana	cactus	cact
Crassula aquatica	water pigmy-weed	cras
Crassula vaginata		cras
Cuscuta gronovii	dodder	cusc
Cynodon aethiopicus	couch grass	poac
Cynodon dactylon	Bermuda grass	poac
Cyrtosperma chamissonis	giant swamp taro	arac
Dactylis glomerata	orchard grass	poac
Daucus carota	wild carrot	apia
Delosperma cooperi	rosy iceplant	mese
Dentaria lasiniata	toothwort	bras
Deschampsia antarctica	antarctic hairgrass	poac
Deuterocohnia chrysantha	bromeliad	brom
Diapensia lapponica	diapensia	diap
Dieffenbachia spp.	dumb cane	arac
Dioscorea elephantipes	wild yam	dios
Distichis spicata	salt grass	poac
Downingia		camp
Draba brachycarpa	short-fruited whitlow-grass	bras
Draba chionophila		bras
Dryas integrifolia	arctic dryad	rosa
Dryas octopetala	mountain dryad	rosa
Dupontia fisheri	grass	poac
Echinochloa crus-galli	barnyard grass	poac
Eichhornia crassipes	water hyacinth	pont
Elatine	waterwort	elat
Eleocharis	spike-rush	cype
Elymus arenarius	lymegrass	poac
Elymus mollis	dune wildrye	poac
Ephedra breana	joint-fir	ephe
Equisetum spp.	horsetails	equi
Eragrostis dikuluwensis	love grass	poac
Eremocarpus setigerus	dove weed	euph
Eriophorum triste	cottongrass	cype
Erodium spp.	stork's bill	gera

Eryngium spp.	coyote thistle	apia
Eschscholzia mexicana	gold poppy	papa
Espeletia spp.	frailejones	aste
Eucalyptus spp.	gum tree	myrt
Eulychnia iquiquensis	cactus	cact
Euphorbia antisyphilitica	candelilla	euph
Euphorbia lactiflua	spurge	euph
Festuca rubra	red fescue	poac
Festuca tolucensis	fescue	poac
Galanthus nivalis	snowdrop	amar
Gentiana urnula	gentian	gent
Geum rossi	alpine avens	rosa
Glaux maritima	sea-milkwort	prim
Halophila engelmanii	Gulf halophila	hydr
Haumaniastrum katangense		lami
Hechtia		brom
Helichrysum newii	strawflower	aste
Heliotropium curassavicum	seaside heliotrope	bora
Heliotropium pycnophyllum	heliotrope	bora
Heliotropium ramosissimum	heliotrope	bora
Helleborus niger	Christmas rose	ranu
Helleborus orientalis	Lenten rose	ranu
Hepatica acutiloba	liverleaf	brass
Hippuris vulgaris	mare's-tail	hipp
Honkenya peploides	sea-beach sandwort	cary
Hydnophytum formicarium		rubi
Ipomoea pes-caprae	railroad vine	conv
Ipomoea stolonifera	white morning glory	conv
Iris germanica	bearded iris	irid
Iris pseudacorus	yellow flag iris	irid
Iris versicolor	blue flag iris	irid
Isoetes	quillwort	isoe
Isopyrum biternatum	false rue-anemone	ranu
Iva imbricata	seashore-elder	aste
Jatropha		euph
Juncus cooperi	Cooper's rush	junc
Kobresia myosuroides	sedge	cype

Lamium amplexicaule	henbit	lami
Lamium hybridum	hybrid dead nettle	lami
Lamium purpureum	red dead nettle	lami
Larrea cuneifolia	compass plant	zygo
Larrea divaricata	jarilla	zygo
Larrea tridentata	creosote bush	zygo
Leontopodium pussilum	Himalayan edelweiss	aste
Leptocarydion vulpiastrum		poac
Lewisia	lewisia	port
Lilaea scilloides	flowering quillwort	jung
Loasa fruticosa		loas
Lobelia cardinalis	cardinal flower	camp
Lobelia telekii	giant lobelia	camp
Lupinus arcticus	arctic lupine	faba
Luzula arctica	arctic woodrush	junc
Luzula confusa	woodrush	junc
Lychnis alpina	copper flower	cary
Lycium deserti	box thorn	sola
Lygodium palmatum	climbing fern	schi
Marsilea vestita	water fern	mars
Medicago lupulina	black medic	faba
Melandrium spp.	bladder campions	cary
Mertensia maritima	sea lungwort	bora
Mesembryanthemum crystallinum	crystalline iceplant	mese
Minuartia arctica	arctic sandwort	cary
Minuartia rubella	boreal sandwort	cary
Monadenium		euph
Myriophyllum	water-milfoil	halo
Nolana lycioides		nola
Nolana peruviana		nola
Nothofagus spp.	southern beeches	faga
Nyssa biflora	swamp tupelo	nyss
Nyssa sylvatica	black tupelo	nyss
Opuntia ficus-indica	Indian fig	cact
Opuntia fragilis	fragile opuntia	cact
Opuntia humifusa	eastern prickly pear	cact
Opuntia pusilla	devil-joint	cact
Orostachys		cras
Oryza sativa	rice	poac

Oxyria digyna	mountain sorrel	poly
Oxytropis kobukensis	milk-vetch	faba
Panicum amarum	bitter panicum	poac
Papaver radicatum	arctic poppy	papa
Paronychia erecta	whitlow-wort	cary
Parrya languinosa	parrya	bras
Paulownia tomentosa	empress tree	bign
Pedicularis lanata	woolly lousewort	scro
Pedilanthus macrocarpus	candelilla	euph
Peperomia	peperomia	pipe
Phippsia algida	snowgrass	poac
Phragmites	common reed	poac
Picea glauca	white spruce	pina
Picea mariana	black spruce	pina
Pilularia americana	water fern	mars
Pinus canariensis	Canary Island pine	pina
Pinus contorta	lodgepole pine	pina
Pinus flexilis	limber pine	pina
Pinus ponderosa	ponderosa pine	pina
Plantago eriopoda	plantain	plan
Plantago maritima	seaside plantain	plan
Poa annua	annual bluegrass	poac
Poa arctica	arctic bluegrass	poac
Polygonum viviparum	viviparous knotweed	poly
Polytrichum	haircap moss	Bryo
Populus tremuloides	aspen	sali
Potamogeton richardsonii	Richardson's pondweed	pota
Potentilla bifurca	cinquefoil	rosa
Potentilla egedii	cinquefoil	rosa
Potentilla hyparctica	arctic cinquefoil	rosa
Potentilla multifida	cinquefoil	rosa
Prosopis juliflora		faba
Prosopis tamarugo	tamarugo	faba
Pseudotsuga menziesii	Douglas fir	pina
Psilocarphus	woolly-heads	aste
Pteris vittata	ladder brake	pter
Puccinellia nuttalliana	Nuttall's alkali grass	poac
Puccinellia phyryganodes	alkali grass	poac
Puccinellia trifolia	alkali grass	poac
Puya boliviensis		brom
Puya raimondii	giant puya	brom

Ranunculus glacialis	glacier crowfoot	ranu
Rendlia cupricola		poac
Racomitrium lanuginosum	hoary rock moss	Bryo
Rhizophora mangle	(Atlantic) red mangrove	rhiz
Rhizophora stylosa		rhiz
Rhodiola	king's crown	cras
Rhus aromatica	fragrant sumac	anac
Ribes nigrum	currant	gros
Ruppia maritima	ditch-grass	pota
Salicornia europaea	European glasswort	chen
Salicornia rubra	annual glasswort	chen
Salicornia subterminalis	perennial glasswort	chen
Salix alaxensis	feltleaf willow	sali
Salix arctica	arctic willow	sali
Salix ovalifolia	ovalleaf willow	sali
Salix pseudopolaris	polar willow	sali
Salix reticulata	netleaf willow	sali
Sassafras albidum	sassafras	sass
Sauromatum guttatum	voodoo lily	arac
Saussurea gossypiphora	cottony saussurea	aste
Saussurea obvallata	Brahma Kamal	aste
Saxifraga cernua	nodding saxifrage	saxi
Saxifraga flagellaris	spiderplant	saxi
Saxifraga foliosa	grained saxifrage	saxi
Saxifraga oppositifolia	purple saxifrage	saxi
Schizachyrium maritimum	seacoast bluestem	poac
Schizachyrium scoparium	little bluestem	poac
Scilla dimartinoi	scilla	lili
Scilla pauciflora	scilla	lili
Scilla siberica	Siberian squill	lili
Scilla socialis	scilla	lili
Scirpus americanus	swordgrass	cype
Scirpus lacustris	common bulrush	cype
Scirpus maritimus	saltwater bulrush	cype
Scirpus paludosus	bulrush	cype
Senecio spp.	groundsels	aste
Senecio congestus	mastodon flower	aste
Senecio meyeri-johannis	groundsel	aste
Senecio telekii	giant groundsel	aste
Senecio vulgaris	common groundsel	aste
Sesuvium portulacastrum	sea purslane	aizo
Silene acaulis	moss campion	cary

Silene cucubalus	bladder campion	cary
Solidago pinetorum	goldenrod	aste
Spergularia canadensis	Canadian sand spurrey	cary
Stellaria crassipes	chickweed	cary
Stellaria decumbens	starwort	cary
Stellaria humifusa	starwort	cary
Stellaria longipes	long-stalked starwort	cary
Stellaria media	common chickweed	cary
Stellaria umbellata	umbrella starwort	cary
Stipa glareosa	needlegrass	poac
Stipa jehu	needlegrass	poac
Suaeda calceoliformis	horned sea-blite	chen
Symplocarpus foetidus	skunk cabbage	arac
Tamarix spp.	tamarisks	tama
Taxodium distichum	bald cypress	taxo
Tephrocactus floccosa	cactus	cact
Tetragonia angustifolia		aizo
Thlaspi rotundifolium	penny-cress	bras
Tillandsia circinnata	air-plant	brom
Tillandsia landbeckii	air-plant	brom
Tradescantia spathacea	Moses-in-the-cradle	comm
Triglochin concinna	arrow-grass	jung
Triglochin maritima	seaside arrow-grass	jung
Typha latifolia	cattail	typh
Uniola paniculata	sea oats	poac
Valeriana	valerian	vale
Viola calaminaria	violet	viol
Viola lutea	violet	viol
Welwitschia mirabilis	welwitschia	welw
Zygophyllum qatarense		zygo

Plant Species Mentioned

Common Name	Latin Name	Taxon
air-plant	*Tillandsia circinnata*	brom
air-plant	*Tillandsia landbeckii*	brom
alyssum	*Alyssum ovirense*	bras
alyssum	*Alyssum wulfenianum*	bras
araucaria	*Araucaria*	arau
arrow-grass	*Triglochin concinna*	jung
arrow-grass, seaside	*Triglochin maritima*	jung
artemisia	*Artemisia herba-alba*	aste
artemisia	*Artemisia skorniakowii*	aste
aspen	*Populus tremuloides*	sali
atriplex, Nuttall's	*Atriplex nuttallii*	chen
Australian pine	*Casuarina*	casu
avens, alpine	*Geum rossi*	rosa
bald cypress	*Taxodium distichum*	taxo
beachgrass, American	*Ammophila breviligulata*	poac
beachgrass, European	*Ammophila arenaria*	poac
bitter-cress	*Cardamine*	bras
bluegrass, annual	*Poa annua*	poac
bluegrass, arctic	*Poa arctica*	poac
bluestem, little	*Schizachyrium scoparium*	poac
bluestem, seacoast	*Schizachyrium maritimum*	poac
box thorn	*Lycium deserti*	sola
Brahma Kamal	*Saussurea obvallata*	aste
bromeliad	*Deuterocohnia chrysantha*	brom
bulrush	*Scirpus paludosus*	cype
bulrush, common	*Scirpus lacustris*	cype
bulrush, saltwater	*Scirpus maritimus*	cype
bursage, white	*Ambrosia dumosa*	aste
cactus	*Copiapoa cernua*	cact
cactus	*Copiapoa haseltoniana*	cact
cactus	*Eulychnia iquiquensis*	cact
cactus	*Tephrocactus floccosa*	cact
calceolaria	*Calceolaria*	scro
calla, wild	*Calla palustris*	arac
campion, bladder	*Silene cucubalus*	cary
campion, moss	*Silene acaulis*	cary
campions, bladder	*Melandrium* spp.	cary
candelilla	*Euphorbia antisyphilitica*	euph
candelilla	*Pedilanthus macrocarpus*	euph

cardinal flower	*Lobelia cardinalis*	camp
carrot, wild	*Daucus carota*	apia
cattail	*Typha latifolia*	typh
chickweed	*Stellaria crassipes*	cary
chickweed, common	*Stellaria media*	cary
Christmas rose	*Helleborus niger*	ranu
cinquefoil	*Potentilla bifurca*	rosa
cinquefoil	*Potentilla egedii*	rosa
cinquefoil	*Potentilla multifida*	rosa
cinquefoil, arctic	*Potentilla hyparctica*	rosa
compass plant	*Larrea cuneifolia*	zygo
copper flower	*Becium homblei*	lami
copper flower	*Lychnis alpina*	cary
cottongrass	*Eriophorum triste*	cype
cowbane	*Cicuta virosa*	apia
coyote thistle	*Eryngium*	apia
creosote bush	*Larrea tridentata*	zygo
crowfoot, glacier	*Ranunculus glacialis*	ranu
currant	*Ribes nigrum*	gros
dead nettle, hybrid	*Lamium hybridum*	lami
dead nettle, red	*Lamium purpureum*	lami
devil-joint	*Opuntia pusilla*	cact
diapensia	*Diapensia lapponica*	diap
ditch-grass	*Ruppia maritima*	pota
dodder	*Cuscuta gronovii*	cusc
Douglas fir	*Pseudotsuga menziesii*	pina
dove weed	*Eremocarpus setigerus*	euph
dryad, arctic	*Dryas integrifolia*	rosa
dryad, mountain	*Dryas octopetala*	rosa
edelweiss, Himalayan	*Leontopodium pussilum*	aste
empress tree	*Paulownia tomentosa*	bign
false rue-anemone	*Isopyrum biternatum*	ranu
fern, ladder brake	*Pteris vittata*	pter
fern, climbing	*Lygodium palmatum*	schi
fern, mosquito	*Azolla pinnata*	azol
fern, water	*Marsilea vestita*	mars
fern, water	*Pilularia americana*	mars
fescue	*Festuca tolucensis*	poac
fescue, red	*Festuca rubra*	poac
Florida rosemary	*Ceratiola ericoides*	empe

foxtail, alpine	*Alopecurus alpinus*	poac
frailejones	*Espeletia* spp.	aste
gentian	*Gentiana urnula*	gent
glasswort, annual	*Salicornia rubra*	chen
glasswort , European	*Salicornia europaea*	chen
glasswort, perennial	*Salicornia subterminalis*	chen
glory-of-the-snow	*Chionodoxa*	lili
goldenrod	*Solidago pinetorum*	aste
goosefoot	*Chenopodium salinum*	chen
goosefoot, red	*Chenopodium rubrum*	chen
grass	*Cynodon aethiopicus*	poac
grass	*Dupontia fisheri*	poac
grass, alkali	*Puccinellia phyryganodes*	poac
grass, alkali	*Puccinellia trifolia*	poac
grass, barnyard	*Echinochloa crus-galli*	poac
grass, bent	*Agrostis tenuis*	poac
grass, Bermuda	*Cynodon dactylon*	poac
grass, creeping bent	*Agrostis stolonifera*	poac
grass, love	*Eragrostis dikuluwensis*	poac
grass, Nuttall's alkali	*Puccinellia nuttalliana*	poac
grass, orchard	*Dactylis glomerata*	poac
grass, perennial bent	*Agrostis perennans*	poac
grass, reed	*Calamagrostis deschampsioides*	poac
grass, reed	*Calamagrostis tolucensis*	poac
grass, rhodes	*Chloris gayana*	poac
grass, salt	*Distichis spicata*	poac
groundsel	*Senecio meyeri-johannis*	aste
groundsel, common	*Senecio vulgaris*	aste
groundsel, giant	*Senecio telekii*	aste
gum tree	*Eucalyptus*	myrt
halophila, Gulf	*Halophila engelmanii*	hydr
hairgrass, antarctic	*Deschampsia antarctica*	poac
heliotrope	*Heliotropium pycnophyllum*	bora
heliotrope	*Heliotropium ramosissimum*	bora
heliotrope, seaside	*Heliotropium curassavicum*	bora
henbit	*Lamium amplexicaule*	lami
Himalayan mustard	*Chorispora*	bras
horsetails	*Equisetum* spp.	equi
Hottentot fig	*Carpobrotus edulis*	mese

iceplant, crystalline	*Mesembryanthemum crystallinum*	mese
iceplant, rosy	*Delosperma cooperi*	mese
Indian fig	*Opuntia ficus-indica*	cact
iodine bush	*Allenrolfea occidentalis*	chen
iris, bearded	*Iris germanica*	irid
iris, blue flag	*Iris versicolor*	irid
iris, yellow flag	*Iris pseudacorus*	irid
jarilla	*Larrea divaricata*	zygo
joint-fir	*Ephedra breana*	ephe
king's crown	*Rhodiola*	cras
knotweed, viviparous	*Polygonum viviparum*	poly
Lenten rose	*Helleborus orientalis*	ranu
lewisia	*Lewisia*	port
liverleaf	*Hepatica acutiloba*	bras
lobelia, giant	*Lobelia telekii*	camp
lousewort, woolly	*Pedicularis lanata*	scro
lungwort, sea	*Mertensia maritima*	bora
lupine, arctic	*Lupinus arcticus*	faba
lymegrass	*Elymus arenarius*	poac
mangrove, black	*Avicennia germinans*	verb
mangrove, (Atlantic) red	*Rhizophora mangle*	rhiz
maple, red	*Acer rubrum*	acer
mare's-tail	*Hippuris vulgaris*	hipp
marsh marigold	*Caltha palustris*	ranu
mastodon flower	*Senecio congestus*	aste
medic, black	*Medicago lupulina*	faba
milk-vetch	*Oxytropis kobukensis*	faba
miner's lettuce	*Claytonia perfoliata*	port
morning glory, white	*Ipomoea stolonifera*	conv
Moses-in-the-cradle	*Tradescantia spathacea*	comm
moss, haircap	*Polytrichum*	Bryo
moss, hoary rock	*Racomitrium lanuginosum*	Bryo
mountain sorrel	*Oxyria digyna*	poly
mouse-ear cress	*Arabidopsis thalinum*	bras
needlegrass	*Stipa glareosa*	poac
needlegrass	*Stipa jehu*	poac
opuntia, fragile	*Opuntia fragilis*	cact

panicum, bitter	*Panicum amarum*	poac
parrya	*Parrya languinosa*	bras
pearlwort	*Colobanthus quitensis*	cary
penny-cress	*Thlaspi rotundifolium*	bras
peperomia	*Peperomia*	pipe
pine, Canary Island	*Pinus canariensis*	pina
pine, limber	*Pinus flexilis*	pina
pine, lodgepole	*Pinus contorta*	pina
pine, ponderosa	*Pinus ponderosa*	pina
plantain	*Plantago eriopoda*	plan
plantain, seaside	*Plantago maritima*	plan
pondweed, Richardson's	*Potamogeton richardsonii*	pota
poppy, arctic	*Papaver radicatum*	papa
poppy, gold	*Eschscholzia mexicana*	papa
prickly pear, eastern	*Opuntia humifusa*	cact
puya, giant	*Puya raimondii*	brom
quillwort	*Isoetes*	isoe
quillwort, flowering	*Lilaea scilloides*	jung
railroad vine	*Ipomoea pes-caprae*	conv
reed, common	*Phragmites*	poac
rice	*Oryza sativa*	poac
rockcress, alpine	*Arabis alpina*	bras
rock jasmine	*Androsace chamaejasme*	prim
rush, Cooper's	*Juncus cooperi*	junc
saltbush	*Atriplex hastata*	chen
saltbush, four-winged	*Atriplex confertifolia*	chen
saltwort	*Batis maritima*	bata
sand spurrey, Canadian	*Spergularia canadensis*	cary
sandwort	*Arenaria patula*	cary
sandwort, arctic	*Minuartia arctica*	cary
sandwort, boreal	*Minuartia rubella*	cary
sandwort, sea-beach	*Honkenya peploides*	cary
sassafras	*Sassafras albidum*	sass
saussurea, cottony	*Saussurea gossypiphora*	aste
saxifrage, grained	*Saxifraga foliosa*	saxi
saxifrage, nodding	*Saxifraga cernua*	saxi
saxifrage, purple	*Saxifraga oppositifolia*	saxi
scilla	*Scilla dimartinoi*	lili
scilla	*Scilla pauciflora*	lili
scilla	*Scilla socialis*	lili

scurvy grass	*Cochlearia*	bras
sea-blite, horned	*Suaeda calceoliformis*	chen
sea-milkwort	*Glaux maritima*	prim
sea oats	*Uniola paniculata*	poac
sea-pink	*Armeria maritima* var. *halleri*	plum
sea purslane	*Sesuvium portulacastrum*	aizo
sea rocket	*Cakile edentula*	bras
seashore-elder	*Iva imbricata*	aste
sedge	*Carex microchaeta*	cype
sedge	*Carex moorcroftii*	cype
sedge	*Carex pachystylis*	cype
sedge	*Carex stans*	cype
sedge	*Carex subspatheca*	cype
sedge	*Carex ursina*	cype
sedge	*Kobresia myosuroides*	cype
sedge, Lyngby's	*Carex lyngbyei*	cype
Siberian squill	*Scilla siberica*	lili
silversword	*Argyroxiphium sandwicense*	aste
skunk cabbage	*Symplocarpus foetidus*	arac
snowdrop	*Galanthus nivalis*	amar
snowgrass	*Phippsia algida*	poac
southern beeches	*Nothofagus* spp.	faga
spiderplant	*Saxifraga flagellaris*	saxi
spike-rush	*Eleocharis*	cype
spring beauty	*Claytonia virginica*	port
spruce, black	*Picea mariana*	pina
spruce, white	*Picea glauca*	pina
spurge	*Euphorbia lactiflua*	euph
starwort	*Stellaria decumbens*	cary
starwort	*Stellaria humifusa*	cary
starwort, long-stalked	*Stellaria longipes*	cary
starwort, umbrella	*Stellaria umbellata*	cary
stork's bill	*Erodium*	gera
strawflower	*Helichrysum newii*	aste
sumac, fragrant	*Rhus aromatica*	anac
swamp taro, giant	*Cyrtosperma chamissonis*	arac
sweet flag	*Acorus calamus*	arac
swordgrass	*Scirpus americanus*	cype
tamarisks	*Tamarix* spp.	tama
tamarugo	*Prosopis tamarugo*	faba
three-seeded mercury	*Acalypha dikuluwensis*	euph
toothwort	*Dentaria lasiniata*	bras

tupelo, black	*Nyssa sylvatica*	nyss
tupelo, swamp	*Nyssa biflora*	nyss
valerian	*Valeriana*	vale
violet	*Viola calaminaria*	viol
violet	*Viola lutea*	viol
voodoo lily	*Sauromatum guttatum*	arac
watergrass	*Bulbostylis mucronata*	cype
water hyacinth	*Eichhornia crassipes*	pont
water-milfoil	*Myriophyllum*	halo
water pigmy-weed	*Crassula aquatica*	cras
water-starwort	*Callitriche*	call
waterwort	*Elatine*	elat
welwitschia	*Welwitschia mirabilis*	welw
white heather, arctic	*Cassiope tetragona*	eric
whitlow-grass, short-fruited	*Draba brachycarpa*	bras
whitlow-wort	*Paronychia erecta*	cary
wildrye, dune	*Elymus mollis*	poac
willow, arctic	*Salix arctica*	sali
willow, feltleaf	*Salix alaxensis*	sali
willow, net-leaf	*Salix reticulata*	sali
willow, ovalleaf	*Salix ovalifolia*	sali
willow, polar	*Salix pseudopolaris*	sali
winterfat	*Ceratoides lanata*	chen
woodrush	*Luzula confusa*	junc
woodrush, arctic	*Luzula arctica*	junc
woolly-heads	*Psilocarphus*	aste
yam, wild	*Dioscorea elephantipes*	dios
yareta	*Azorella*	apia

Animal Species Mentioned

Latin Name	Common Name	Taxon
Actornithophilus patellatus	curlew quill louse	Mall
Antarctophthirus ogmorhini	Weddell seal louse	Mall
Antidorcas marsupialis	springbok	bovi
Aptenodytes forsteri	emperor penguin	sphe
Aptenodytes patagonicus	king penguin	sphe
Araneus diadematus	European garden spider	argi
Archistoma besselsi	springtail	Coll
Arenivaga investigata	desert cockroach	blat
Arixenia esau	earwig	derm
Arixenia jacobsoni	earwig	derm
Artemia salina	brine shrimp	Anos
Attagenus megatoma	black carpet beetle	derm
Belgica antarctica	antarctic midge	chir
Bos grunniens	yak	bovi
Bradypodicola spp.	three-toed-sloth moth	pyra
Bradypus	three-toed sloth	brad
Camelus dromedarius	Arabian camel	came
Carduelis flammea	common redpoll	frin
Carnus hemapterus	fly	mili
Catostomus catostomus	longnosed sucker	cato
Cediopsylla simplex	cottontail rabbit flea	Siph
Cenocorixa bifida	water boatman	cori
Cenocorixa expleta	water boatman	cori
Chionea	snow-fly	tipu
Chionis alba	American sheathbill	chio
Chironomus spp.	chironomid midges	chir
Chlorohydra viridissima	green hydra	hydi
Cicindela marutha	tiger beetle	cici
Conicera tibialis	coffin fly	phor
Convoluta roscoffensis	flatworm	Plat
Coregonus	whitefish	salm
Coxiella spp.	salt lake snails	Gast
Cricotopus variabilis	chironomid midge	chir
Cyprinodon spp.	killifish	cypo
Cyprinodon milleri	Cottonball Marsh pupfish	cypo
Dasyhelea thompsoni	biting midge	cera
Delia radicum	cabbage root fly	musc
Diamesa zernyi	chironomid midge	chir

Dicrostonyx spp.	collared lemming	cric
Donax spp.	coquina clam	dona
Echidnophaga gallinacea	poultry sticktight flea	tung
Emerita portoricensis	mole crab	hipp
Enochrus diffusus	water scavenger beetle	hydr
Ephydra cinerea	brine fly	ephy
Ephydra hians	brine fly	ephy
Euchondrus albulus	snail	Gast
Euchondrus desertorum	snail	Gast
Euchondrus ramonensis	snail	Gast
Eunice gigantea	sandworm	euni
Eurosta solidaginis	goldenrod gall fly	teph
Gasterophilus haemorroidalis	horse bot fly	gast
Gasterophilus intestinalis	horse bot fly	gast
Gasterophilus nigricornis	horse bot fly	gast
Gasterophilus precorum	horse bot fly	gast
Gecarcinus lateralis	red land crab	geca
Geococcyx californianus	roadrunner	cucu
Grus americana	whooping crane	grui
Grylloblatta spp.	rock crawlers	gryl
Gynaephora groenlandica	woolly bear caterpillar	lipa
Gyrostigma	rhinoceros bot fly	gast
Haloniscus searlei	salt lake slater	Isop
Heleomyza borealis	heleomyzid fly	hele
Heminirus	earwig	derm
Hyalella azteca	fresh-water amphipod	Amph
Hymenolepis nana	dwarf tapeworm	Cest
Iridomyrmex cordatus	ant	form
Isotoma spp.	springtails	Coll
Julus spp.	snake millipedes	Juli
Lagopus lagopus	willow ptarmigan	tetr
Lagopus mutus	rock ptarmigan	tetr
Lama guanicoe	guanaco	came
Lancetes lanceolatus	predaceous diving beetle	dyti
Lasioderma serricorne	cigarette beetle	anob
Laurus dominicanus	southern black-backed gull	lari
Lemmus sibiricus	brown lemming	cric

Lepidophthirus macrorhini	southern elephant seal louse	Mall
Lepisma	silverfish	lepi
Leptonychotes weddelli	Weddell seal	phoc
Lepus arcticus	arctic hare	lepo
Lerwa lerwa	snow partridge	phas
Limulus polyphemus	Atlantic horseshoe crab	Mero
Machilanus spp.	glacier fleas	mach
Machilis fuscistylis	jumping bristletail	mach
Mansonia richerdii	mosquito	culi
Megaselia scalaris	humpbacked fly	phor
Mustela altaica	Tibetan weasel	must
Mustela erminea	ermine	must
Mysis oculata relicta	opossum shrimp	Mysi
Nacella concinna	antarctic limpet	Gast
Nosopsyllus fasciatus	rat flea	Siph
Ocypode quadrata	ghost crab	ocyp
Odontopsyllus multispinosus	cottontail rabbit flea	Siph
Onthophagus	dung beetle	scar
Ornithomyia fringillina	louse fly	hipo
Ovibos moschatus	muskox	bovi
Ovis amon hodgsoni	Tibetan argali	bovi
Pagodroma nivea	snow petrel	proc
Pagophilus groenlandicus	harp seal	phoc
Pantholops hodgsoni	chiru	bovi
Parapsyllus magellanicus	rockhopper penguin flea	Siph
Parartemia	Australian brine shrimp	Anos
Pedicia hannai	crane fly	tipu
Phoeniconaias	lesser flamingo	phoe
Phoenicopterus	greater flamingo	phoe
Phylodecta laticollis	leaf beetle	chry
Piophila casei	cheeseskipper fly	piop
Placobranchus	nudibranch	Nudi
Platicobboldia loxodontis	elephant bot	gast
Polydesmus spp.	flat-backed millipedes	Poly
Polypedilum vanderplankii	chironomid midge	chir
Pseudois nayaur	bharal or bluesheep	bovi
Ptinus tectus	spider beetle	ptin

Rana sylvatica	wood frog	rani
Rangifer tarandus	caribou and reindeer	cerv
Rangifer tarandus pearyi	Peary's caribou	cerv
Rasbora sp.	rasbora	cypr
Rhantus pulverosus	predaceous diving beetle	dyti
Rhyniella praecursor	extinct springtail	Coll
Sarcophaga crassipalpis	flesh fly	sarc
Scaphiopus spp.	spadefoot toads	pelo
Scaphiopus couchi	Couch's spadefoot toad	pelo
Siren intermedia	lesser siren	sire
Siren lacertina	greater siren	sire
Spilopsyllus cuniculi	rabbit flea	Siph
Stegobium paniceum	drugstore beetle	anob
Sturnidoecus sturni	starling louse	Mall
Tanytarsus barbitaris	chironomid midge	chir
Taurotragus oryx	eland	bovi
Tenebrio molitor	yellow meal worm	tene
Thermobia domestica	firebrat	lepi
Tribolium spp.	flour beetle	tene
Trichocorixa verticalis	water boatman	cori
Trichocorixa verticalis interiores	water boatman	cori
Tridacna	giant clam	trid
Trochiloecetes	bird louse	Mall
Tryonia	snail	Gast
Tubifex spp.	sewage worms	tubi
Tunga penetrans	chigoe	tung
Vertagopus arcticus	springtail	Coll

Animal Species Mentioned

Common Name	Latin Name	Taxon
amphipod, fresh-water	*Hyalella azteca*	Amph
ant	*Iridomyrmex cordatus*	form
argali, Tibetan	*Ovis amon hodgsoni*	bovi
beetle, black carpet	*Attagenus megatoma*	derm
beetle, cigarette	*Lasioderma serricorne*	anob
beetle, drugstore	*Stegobium paniceum*	anob
beetle, dung	*Onthophagus*	scar
beetle, flour	*Tribolium*	tene
beetle, leaf	*Phylodecta laticollis*	chry
beetle, predaceous diving	*Rhantus pulverosus*	dyti
beetle, predaceous diving	*Lancetes lanceolatus*	dyti
beetle, spider	*Ptinus tectus*	ptin
beetle, tiger	*Cicindela marutha*	cici
beetle, water scavenger	*Enochrus diffusus*	hydr
bharal or bluesheep	*Pseudois nayaur*	bovi
biting midge	*Dasyhelea thompsoni*	cera
brine shrimp	*Artemia salina*	Anos
brine shrimp, Australian	*Parartemia*	Anos
bristletail, jumping	*Machilis fuscistylis*	mach
camel, Arabian	*Camelus dromedarius*	came
caribou and reindeer	*Rangifer tarandus*	cerv
caribou, Peary's	*Rangifer tarandus pearyi*	cerv
chigoe	*Tunga penetrans*	tung
chiru	*Pantholops hodgsoni*	bovi
clam, coquina	*Donax* spp.	dona
clam, giant	*Tridacna*	trid
cockroach, desert	*Arenivaga investigata*	blat
crab, ghost	*Ocypode quadrata*	ocyp
crab, mole	*Emerita portoricensis*	hipp
crab, red land	*Gecarcinus lateralis*	geca
crane fly	*Pedicia hannai*	tipu
crane, whooping	*Grus americana*	grui
earwig	*Arixenia esau*	Derm
earwig	*Arixenia jacobsoni*	Derm
earwig	*Heminirus*	Derm
eland	*Taurotragus oryx*	bovi
ermine	*Mustela erminea*	must

firebrat	*Thermobia domestica*	lepi
flamingo, greater	*Phoenicopterus*	phoe
flamingo, lesser	*Phoeniconaias*	phoe
flatworm	*Convoluta roscoffensis*	Plat
flea, cottontail rabbit	*Odontopsyllus multispinosus*	Siph
flea, cottontail rabbit	*Cediopsylla simplex*	Siph
flea, poultry	*Echidnophaga gallinacea*	tung
flea, rabbit	*Spilopsyllus cuniculi*	Siph
flea, rat	*Nosopsyllus fasciatus*	Siph
flea, rockhopper penguin	*Parapsyllus magellanicus*	Siph
fly	*Carnus hemapterus*	mili
fly, heleomyzid	*Heleomyza borealis*	hele
fly, brine	*Ephydra cinerea*	ephy
fly, brine	*Ephydra hians*	ephy
fly, cabbage root	*Delia radicum*	musc
fly, cheeseskipper	*Piophila casei*	piop
fly, coffin	*Conicera tibialis*	phor
fly, elephant bot	*Platicobboldia loxodontis*	gast
fly, flesh	*Sarcophaga crassipalpis*	sarc
fly, goldenrod gall	*Eurosta solidaginis*	teph
fly, horse bot	*Gasterophilus haemorroidalis*	gast
fly, horse bot	*Gasterophilus intestinalis*	gast
fly, horse bot	*Gasterophilus nigricornis*	gast
fly, horse bot	*Gasterophilus precorum*	gast
fly, humpbacked	*Megaselia scalaris*	phor
fly, louse	*Ornithomyia fringillina*	hipo
fly, rhinoceros bot	*Gyrostigma*	gast
frog, wood	*Rana sylvatica*	rani
glacier fleas	*Machilanus* spp.	mach
guanaco	*Lama guanicoe*	came
gull, southern black-backed	*Laurus dominicanus*	lari
hare, arctic	*Lepus arcticus*	lepo
harp seal	*Pagophilus groenlandicus*	phoc
horseshoe crab, Atlantic	*Limulus polyphemus*	Mero
hydra, green	*Chlorohydra viridissima*	hydi
killifish	*Cyprinodon* spp.	cypo
lemming, brown	*Lemmus sibiricus*	cric
lemming, collared	*Dicrostonyx* spp.	cric
limpet, antarctic	*Nacella concinna*	Gast

long-nose sucker	*Catostomus catostomus*	cato
louse, bird	*Trochiloecetes*	Mall
louse, curlew quill	*Actornithophilus patellatus*	Mall
louse, southern elephant seal	*Lepidophthirus macrorhini*	Mall
louse, starling	*Sturnidoecus sturni*	Mall
louse, Weddell seal	*Antarctophthirus ogmorhini*	Mall
meal worm, yellow	*Tenebrio molitor*	tene
midge, antarctic	*Belgica antarctica*	chir
midge, chironomid	*Cricotopus variabilis*	chir
midge, chironomid	*Diamesa zernyi*	chir
midge, chironomid	*Polypedilum vanderplankii*	chir
midge, chironomid	*Tanytarsus barbitaris*	chir
midges, chironomid	*Chironomus* spp.	chir
millipedes, flat-backed	*Polydesmus* spp.	Poly
millipedes, snake	*Julus* spp.	Juli
mosquito	*Mansonia richerdii*	culi
moth, three-toed-sloth	*Bradypodicola* spp.	pyra
muskox	*Ovibos moschatus*	bovi
nudibranch	*Placobranchus*	Nudi
opossum shrimp	*Mysis oculata relicta*	Mysi
partridge, snow	*Lerwa lerwa*	phas
penguin, emperor	*Aptenodytes forsteri*	sphe
penguin, king	*Aptenodytes patagonicus*	sphe
petrel, snow	*Pagodroma nivea*	proc
ptarmigan, rock	*Lagopus mutus*	tetr
ptarmigan, willow	*Lagopus lagopus*	tetr
pupfish, Cottonball Marsh	*Cyprinodon milleri*	cypo
rasbora	*Rasbora* sp.	cypr
redpoll, common	*Carduelis flammea*	frin
roadrunner	*Geococcyx californianus*	cucu
rock crawlers	*Grylloblatta* spp.	gryl
sandworm	*Eunice gigantea*	euni
sewage worms	*Tubifex* spp.	tubi
sheathbill, American	*Chionis alba*	chio
silverfish	*Lepisma*	lepi
siren, greater	*Siren lacertina*	sire
siren, lesser	*Siren intermedia*	sire

slater, salt lake	*Haloniscus searlei*	Isop
sloth, three-toed	*Bradypus*	brad
snail	*Euchondrus albulus*	Gast
snail	*Euchondrus desertorum*	Gast
snail	*Euchondrus ramonensis*	Gast
snail	*Tryonia*	Gast
snails, salt lake	*Coxiella* spp.	Gast
snow-fly	*Chionea*	tipu
spadefoot toad, Couch's	*Scaphiopus couchi*	pelo
spadefoot toads	*Scaphiopus* spp.	pelo
spider, European garden	*Araneus diadematus*	argi
springbok	*Antidorcas marsupialis*	bovi
springtail	*Archistoma besselsi*	Coll
springtail	*Vertagopus arcticus*	Coll
springtail, extinct	*Rhyniella praecursor*	Coll
springtails	*Isotoma* spp.	Coll
tapeworm, dwarf	*Hymenolepis nana*	Cest
water boatman	*Cenocorixa bifida*	cori
water boatman	*Cenocorixa expleta*	cori
water boatman	*Trichocorixa verticalis*	cori
water boatman	*Trichocorixa verticalis interiores*	cori
weasel, Tibetan	*Mustela altaica*	must
whitefish	*Coregonus*	salm
woolly bear caterpillar	*Gynaephora groenlandica*	lipa
yak	*Bos grunniens*	bovi

Taxonomic Codes

Plants

acer	maple family	Aceraceae
aizo	carpetweed family	Aizoaceae
amar	amaranth famiily	Amaranthaceae
anac	cashew family	Anacardiaceae
apia	carrot family	Apiaceae
arac	arum family	Araceae
arau	araucaria family	Araucariaceae
aste	daisy family	Asteraceae
azol	mosquito fern family	Azollaceae
bata	saltwort family	Bataceae
bign	trumpet vine family	Bignoniaceae
bora	borage family	Boraginaceae
bras	mustard family	Brassicaceae
brom	pineapple family	Bromeliaceae
Bryo	mosses	Bryophyta (Division)
cact	cactus family	Cactaceae
call	water-starwort family	Callitrichaceae
camp	bellflower family	Campanulaceae
cary	pink family	Caryophyllaceae
casu	Australian pine family	Cassuarinaceae
chen	goosefoot family	Chenopodiaceae
comm	spiderwort family	Commelinaceae
conv	morning glory family	Convulvulaceae
cras	stonecrop family	Crassulaceae
cusc	dodder family	Cuscutaceae
cype	sedge family	Cyperaceae
diap	diapensia family	Diapensiaceae
dios	yam family	Dioscoriaceae
elat	waterwort family	Elatinaceae
empe	crowberry family	Empetraceae
ephe	mormon tea family	Ephedraceae
equi	horsetail family	Equisetaceae
eric	heath family	Ericaceae
euph	spurge family	Euphorbiaceae
faba	legume family	Fabaceae
faga	oak family	Fagaceae

gent	gentian family	Gentianaceae
gera	geranium family	Geraniaceae
gros	gooseberry family	Grossulariaceae
halo	water-milfoil family	Haloragaceae
hipp	mare's-tail family	Hippuridaceae
hydr	waterweed family	Hydrocharitaceae
irid	iris family	Iridaceae
isoe	quillwort family	Isoetaceae
junc	needle rush family	Juncaceae
jung	arrowgrass family	Juncaginaceae
lami	mint family	Lamiaceae
lili	lily family	Liliaceae
loas	loasa family	Loasaceae
lyco	club-moss family	Lycopodiaceae
mars	Marsilea family	Marsileaceae
mese	iceplant Family	Mesembryanthemaceae
myrt	myrtle family	Myrtaceae
nola	nolana family	Nolanaceae
nyss	tupelo family	Nyssaceae
papa	poppy family	Papaveraceae
pina	pine family	Pinaceae
pipe	pepper family	Piperaceae
plan	plantain family	Plantaginaceae
plum	plumbago family	Plumbaginaceae
poac	grass family	Poaceae
poly	dock family	Polygonaceae
pont	pickerel weed family	Pontederiaceae
port	purselane family	Portulacaceae
pota	pondweed family	Potamogetonaceae
prim	primrose family	Primulaceae
pter	brake fern family	Pteridaceae
ranu	buttercup family	Ranunculaceae
rhiz	red mangrove family	Rhizophoraceae
rosa	rose family	Rosaceae
rubi	madder family	Rubiaceae

sali	willow family	Salicaceae
sass	sassafrass family	Sassafraceae
saxi	saxifrage family	Saxifragaceae
schi	climbing fern family	Schizaeaceae
scro	figwort family	Scrophulariaceae
sola	nightshade family	Solanaceae
tama	tamarisk family	Tamaricaceae
taxo	bald cypress family	Taxodiaceae
typh	cattail family	Typhaceae
vale	valerian Family	Valerianaceae
verb	verbena family	Verbenaceae
viol	violet family	Violaceae
welw	Welwitschia Family	Welwitschiaceae
zygo	caltrop family	Zygophyllaceae

Taxonomic Codes

Animals

Amph	amphipods	Amphipoda (Order)
anob	death-watch beetle fam.	Anobiidae
Anos	fairy shrimps	Anostraca (Order)
argi	orb-weaving spider fam.	Argiopidae
blat	cockroach family	Blattidae
bovi	cattle family	Bovidae
brad	three-toed sloth family	Bradypodidae
came	camel family	Camelidae
cato	carp-like sucker fam.	Catostomidae
Cest	tapeworms	Cestoda (Class)
cera	biting midges	Ceratopogonidae
cerv	deer family	Cervidae
chio	sheathbill family	Chionididae
chir	midge family	Chironomidae
chry	leaf beetle family	Chrysomelidae
cici	tiger beetle family	Cicindelidae
Coll	springtails	Collembola (Order)
cori	water boatman family	Corixidae
cric	gerbil family	Cricetidae
cucu	cuckoo family	Cuculidae
culi	mosquito family	Culicidae
cypo	killifish family	Cyprinodontidae
cypr	carp family	Cyprinidae
Derm	earwigs	Dermaptera (Order)
derm	carpet beetle family	Dermestidae
dona	coquina family	Donacidae
dyti	predaceus diving beetles	Dytiscidae
ephy	shore fly family	Ephydridae
euni	family of polychaetes	Eunicidae
form	ant family	Formicidae
frin	finch family	Fringillidae
gast	bot fly family	Gasterophilidae
Gast	snails and limpets	Gastropoda (Class)
geca	land crab family	Geocarcinidae

grui	crane family	Gruidae
gryl	rock crawler family	Grylloblattidae
hele	heleomyzid fly family	Heleomyzidae
hipo	louse fly family	Hippoboscidae
hipp	mole crab family	Hippidae
hydi	hydra family	Hydridae
hydr	water scavenger beetles	Hydrophilidae
Isop	isopods	Isopoda (Order)
Juli	snake millipedes	Julida (Order)
lari	gull family	Laridae
lepi	silverfish family	Lepismatidae
lepo	rabbit family	Leporidae
lipa	liparid moth family	Liparidae
mach	snow flea family	Machilidae
Mall	lice	Mallophaga (Order)
Mero	horseshoe crabs	Merostomata (Class)
mili	milichiid fly family	Milichiidae
musc	house fly family	Muscidae
must	weasel family	Mustelidae
mysi	opossum shrimp	Mysidacea (Order)
Nudi	nudibranchs	Nudibranchia (Order)
ocyp	ghost crab family	Ocypodidae
pelo	spadefoot family	Pelobatidae
phas	partridge family	Phasianidae
phoc	seal family	Phocidae
phoe	flamingo family	Phoenicopteridae
phor	humpback fly family	Phoridae
piop	skipper fly family	Piophilidae
Plat	flatworms	Platyhelminthes (Phylum)
Poly	sandworms	Polychaeta (Class)
proc	shearwater family	Procellariidae
ptin	spider beetle family	Ptinidae
pyra	pyralid moths	Pyralidae
rani	bullfrog family	Ranidae

salm	salmon family	Salmonidae
sarc	flesh-fly family	Sarcophagidae
scar	dung beetle family	Scarabaeidae
Siph	fleas	Siphonaptera (Order)
sire	siren family	Sirenidae
sphe	penguin family	Spheniscidae
tene	darkling beetle family	Tenebrionidae
teph	fruit fly family	Tephritidae
tetr	grouse family	Tetraonidae
tipu	crane fly family	Tipulidae
trid	giant clam family	Tridaenidae
tubi	sewage worms	Tubificidae
tung	stick-tight flea family	Tungidae

Places Mentioned

Place	Description	Location	Lat	Long
Abert, Lake		Oregon	42.38N	120.13W
Alexandra Fjord		Ellesmere I.	78.88N	75.92W
Altai	Mountains	Asia	48.00N	90.00E
Altar, Desierto de	Desert	Sonora	31.50N	114.15W
Antofagasta	City	Chile	23.39S	70.24W
Arica	City	Chile	18.29S	70.20W
Atacama, Desierto de	Desert	S. A.	24.30S	69.15W
Athabasca River		Can.	58.40N	110.50W
Athabasca, Lake		Can.	59.07N	110.00W
Baffin I.		Canada	68.00N	70.00W
Bahrain	Island	Persian Gulf	26.00N	50.30E
Baja California	Peninsula	Mexico	32.18N	115.12W
Barter I.		Alaska	70.08N	143.35W
Salar de Bellavista	Salt Plain	Chile	33.31S	70.37W
Bering Sea		Artic	60.00N	175.00W
Canning River		Alaska	70.05N	145.30W
Cerro Moreno	Headland	Chile	23.28S	70.25W
Chang Tang	High Plateau	Tibet	32.00N	85.00E
Chaplin	Lake	Sask. Can.	50.18N	106.35W
Chihuahuan Desert		N. A.	35.00N	106.00W
Quebrada La Chimba	Creek	Chile	23.39S	70.24W
Colville River		Alaska	70.25N	150.30W
Copperbelt	District	Zambia	13.00S	28.00E
Coral Pink S. Dunes	State Park	Utah	37.00N	112.50W
Cornwallis I.		Canada	75.15N	94.30W
Death Valley	Nat. Mon.	Cal.	36.30N	117.00W
Dolomite Alps		Austria/Italy	46.25N	11.50E
Ellesmere I.		Canada	81.00N	80.00W
Galápagos Is.	Nat. Park	Ecuador	0.15S	90.15W
Gobi	Desert	Mongolia	43.00N	105.00E
Great Salt Lake		Utah	41.10N	112.30W
Quebrada Huantajaya	Creek	Chile	20.13S	70.10W
Icefield Ranges		Can.	60.50N	140.00W
Iquique	City	Chile	20.13S	70.10W
Isola di Lampedusa	Island	Med. Sea	35.50N	12.50E
Jago River		Alaska	69.50N	143.00W
Kalahari	Semi-desert	S. Africa	24.00S	21.3E
Kamet	Mountain	Himalayas	30.54N	79.37E
Karoo	Region	South Africa	29.00S	18.00E
Kilamanjaro	Mountain	Tanzania	3.04S	37.22E
Kluane Lake		Yukon	61.15N	138.40W

Kluane Nat. Park		Yukon	60.45N	139.30W
Kobuk River		Alaska	66.45N	161.00W
Kobuk Valley	Nat. Park	Alaska	67.20N	159.00W
Kolwezi	Town	Zaire	10.43S	25.28E
Likasi	Town	Zaire	10.59S	26.44E
Lancaster Sound		Canada	74.13N	84.00W
Lehigh Gap		Penn. USA	40.70N	75.36W
Loa River		Chile	21.26S	70.04W
Lubumbashi	City	Zaire	11.40S	27.28E
Mackenzie Delta		Can.	68.50N	135.25W
Mackenzie Mts.		Can.	64.00N	130.00W
Mackenzie River		N.T. Can.	69.15N	134.08W
Magdalena, Llano	Plain	Baja Cal.	24.55N	111.50W
Makalu	Mountain	Himalayas	27.54N	87.06E
Mauna Kea	Volcano	Hawaii	19.50N	155.28W
Mauna Loa	Volcano	Hawaii	19.29N	155.36W
Meade River		Alaska	70.50N	156.25W
Mojave Desert		U. S.	35.00N	117.00W
Mono Lake		Cal.	38.18N	119.22W
Monte	Semi-desert	Argentina	30.00S	65.00W
Muskiki	Lake	Sask. Can.	52.00N	105.00W
Namib Desert		Namibia	23.00S	15.00E
Negev	Desert	Israel	30.30N	34.55E
New Caledonia	Islands	Coral Sea	21.30S	165.30E
New Idria	Hills	Cal.	36.20N	120.40W
Nunivak I.		Alaska	60.00N	166.30W
Oimyakon	Village	Siberia	63.47N	142.82E
Okanagan Mnt.	Prov. Park	Can.	49.45N	119.40W
Orizaba, Pico de	Nat. Park	Mex.	19.01N	97.16W
Palmerton	City	Penn. USA	40.48N	75.36W
Pamir	Mountains	Tajikistan	38.00N	73.00E
Paposo	City	Chile	25.01S	70.28W
Patagonia	Region	S. A.	44.00S	68.00W
Peary Land	Region	Greenland	83.00N	35.00W
Piedmont	Uplands	E. USA	36.00N	81.50W
Queensland	Territory	Australia	22.00S	145.00E
Ross Ice Shelf		Antarctica	81.30S	175.00W
Sahara	Desert	N. Africa	26.00N	13.00E
Sahel	Region	N. Africa	10.00N	20.00E
San Felipe, Des. de	Desert	Baja Cal.	31.00N	114.52W
Sanibel I.		Florida	26.27N	82.06W
Scoresby Sound		Greenland	70.50N	25.00W
Serena, La	City	Chile	29.54S	71.16W

Seward Peninsula		Alaska	65.00N	164.00W
Shaba Province		Zaire	8.00S	27.00E
Snag	Village	Yukon	62.24N	140.22W
Socompa	Volcano	Chile/Argen.	24.42S	68.25W
Sonoran Desert		U.S./Mexico	30.00N	113.00W
South Orkney Is.		G. B.	60.35S	45.30W
St. Elias Mts.		Can.	60.30N	139.30W
Surtsey	Island	Iceland	63.16N	20.32W
Svalbard	Archipelago	Norway	78.00N	20.00E
Tajikistan	Republic	Asia	39.00N	71.00E
Takla Makan	Desert	China	39.00N	83.00E
Tamarugal, Pampa del	Plain	Chile	21.00S	69.25W
Taymyr Peninsula		Russia	76.00N	104.00E
Ungava	Peninsula	Can.	60.00N	74.00W
Victoria Land	Region	Antarctica	75.00S	163.00E
Vizcaíno, El	Desert	Baja Cal.	27.40N	113.40W
Windward Is.		Caribbean	13.00N	61.00W
Wrangel I.		Russia	71.00N	179.30W
Yakutat	Village	Alaska	59.50N	139.00W
Yucatán	Peninsula	Mex.	19.30N	89.00W

Selected References

1 *The Extremes of the Past*

de Duve, C. 1995. The Beginnings of Life on Earth. *American Scientist.* 83: 428-437.

Eldredge, N. 1987. *Life Pulse: Episodes from the Story of the Fossil Record.* New York: Facts on File Publications.

Gould, S. J. 1989. *Wonderful Life: The Burgess Shale and the Nature of History.* New York: W. W. Norton & Company.

Wayne, R. P. 1991. 2nd Ed. *Chemistry of Atmospheres.* Oxford: Clarendon Press.

2 *Aridity and its Effect on Life*

Abushama, F. T. 1984. Epigeal Insects. In Cloudsley-Thompson, J. L., ed. *Sahara Desert*, pp. 129-144. New York: Pergamon Press.

Brown, J. H., O. J. Reichman, and D. W. Davidson. 1979. Granivory in Desert Ecosystems. *Ann. Rev. Ecol. Syst.* 10: 201-227.

Cook, O. F. 1917. Polar Bear Cacti. *Journal of Heredity.* 8: 113.

Cornes, M. D. 1989. *The Wild Flowering Plants of Bahrain.* London: Immel Publishing Limited.

Crosswhite, F. S. and C. D. Crosswhite. 1982. The Sonoran Desert. In Bender, G. L., ed. *Reference Handbook on the Deserts of North America*, pp. 163-319. Westport, CT: Greenwood Press.

DeDecker, M. 1984. *Flora of the Northern Mojave Desert, California.* Berkeley, CA: California Native Plant Society.

Evenari, M., E. D. Schulze, O. L. Lange, L. Kappen, and U. Buschbom. 1976. Plant Production in Arid and Semi-Arid Areas. In O. L. Lange, L. Kappen, and E. D. Schulze, eds. *Water and Plant Life*, pp. 439-451. New York: Springer-Verlag.

Everitt, J. H. and D. L. Drawe. 1993. *Trees, Shrubs & Cacti of South Texas.* Lubbock, TX: Texas Tech University Press.

Fitter, A. H. and R. K. M. Hay. 1987. *Environmental Physiology of Plants.* London: Academic Press, Inc.

Happold, D. C. D. 1984. Small Mammals. In Cloudsley-Thompson, J. L., ed. *Sahara Desert*, pp. 251-275. New York: Pergamon Press.

Ihlenfeldt, H. D. 1994. Diversification in an Arid World: The Mesembryanthemaceae. *Annu. Rev. Ecol. Syst.* 25: 521-546.

Innes, C. and C. Glass. 1991. *Cacti.* New York: Portland House.

Jones, C. G. and M. Shachak. 1994. Desert Snails' Daily Grind. *Natural History* 103: 56-61.

Kassas, M. and K. H. Batanouny. 1984. Plant Ecology. In Cloudsley-Thompson, J. L., ed. *Sahara Desert*, pp. 77-90. New York: Pergamon Press.

Lewis, J. G. E. 1984. Woodlice and Myriapods. In Cloudsley-Thompson, J. L., ed. *Sahara Desert*, pp. 115-127. New York: Pergamon Press.

Lieth, H. 1976. The Use of Correlation Models to Predict Primary Productivity from Precipitation or Evapotranspiration. In O. L. Lange, L. Kappen, and E. D. Schulze, eds. *Water and Plant Life*, pp. 392-407. New York: Springer-Verlag.

Page, J. 1984. *Arid Lands (Planet Earth Series)*. Alexandria, VA: Time-Life Books.

Polunin, N. 1960. *Introduction to Plant Geography and Some Related Sciences*. New York: McGraw-Hill Book Company.

Richter, H. 1976. The Water Status in the Plant Experimental Evidence. In O. L. Lange, L. Kappen, and E. D. Schulze, eds. *Water and Plant Life*, pp. 42-58. New York: Springer-Verlag.

Roberts, N. C. 1989. *Baja California Plant Field Guide*. La Jolla, CA: Natural History Publishing Co.

Rowlands, P., H. Johnson, E. Ritter, and A. Endo. 1982. The Mohave Desert. In Bender, G. L., ed. *Reference Handbook on the Deserts of North America*, pp. 103-162. Westport, CT: Greenwood Press.

Rundel, P. W., M. O. Dillon, B. Palma, H. A. Mooney, S. L. Gulmon, and J. R. Ehleringer. 1991. The Phytogeography and Ecology of the Coastal Atacama and Peruvian Deserts. *ALISO*. 13(1): 1-49.

Sajeva, M. and M. Costanzo. 1994. *Succulents: The Illustrated Dictionary*. Portland, OR: Timber Press.

Stocker, O. 1976. The Water-Photosynthesis Syndrome and the Geographical Plant Distribution in the Saharan Desert. In O. L. Lange, L. Kappen, and E. D. Schulze, eds. *Water and Plant Life*, pp. 506-521. New York: Springer-Verlag.

Takhtajan, A. L. 1986. *Floristic Regions of the World*. Berkeley: University of California Press.

Von Willert, D. J., B. M. Eller, M.J.A. Werger, E. Brinckmann, and H. D. Ihlenfeldt. 1992. *Life Strategies of Succulents in Deserts: with Special Reference to the Namib desert*. New York: Cambridge University Press.

Wallace, J. M. and P. V. Hobbs. 1977. *Atmospheric Science: An Introductory Survey*. Orlando, FL: Academic Press, Inc.

Wickens, G. E. 1984. Flora. In Cloudsley-Thompson, J. L., ed. *Sahara Desert*, pp. 67-75. New York: Pergamon Press.

3 *Your House is My House*

Askew, R. R. 1971. *Parasitic Insects*. New York: American Elsevier Publishing Company.

Burr, M. and K. Jordan. 1912. On *Arixenia* Burr, a suborder of Dermaptera. *Trans. Int. Congr. Ent.* 2: 398-421.

Evans, H. E. 1984. Insect Biology: A Textbook of Entomology. Reading, MA: Addison-Wesley Publishing Company.

Hadley, N. F. 1994a. The Uptake of Water, ch. 6 in *Water Relations of Terrestrial Arthropods*, pp. 222-267. San Diego, CA: Academic Press, Inc.

Hadley, N. F. 1994b. Thermoregulation and Water Relations, ch. 9 in *Water Relations of Terrestrial Arthropods*, pp. 324-344. San Diego, CA: Academic Press, Inc.

Pratt, H. D., K. S. Littig, and H. G. Scott. 1976. *Household and Stored-Food Insects of Public Health Importance and Their Control*. Atlanta, GA: U.S. Department of Health, Education, and Welfare. HEW Publication No. (CDC) 76-8122.

Selkirk, P. M., R. D. Seppelt, and D. R. Selkirk. 1990. Microbiology, Parasitology and Terrestrial Arthropods, ch. 10 in *Subantarctic Macquarie Island: Environment and Biology*, pp. 188-202. New York: Cambridge University Press.

Wigglesworth, V. B. 1972. *The Principles of Insect Physiology*. London: Chapman and Hall.

Wigglesworth, V. B. 1974. *Insect Physiology*. London: Chapman and Hall.

4 *From Sand to Serpentine*

Bache, C. A. and D. J. Lisk. 1990. Heavy-Metal Absorption by Perennial Ryegrass and Swiss Chard Grown in Potted Soils Amended with Ashes from 18 Municipal Refuse Incinerators. *J. Agric. Food. Chem.* 38: 190-194.

Baker, A. J. M. and R. R. Brooks. 1989. Terrestrial Higher Plants which Hyper-accumulate Metallic Elements: A Review of their Distribution, Ecology and Phytochemistry. *Biorecovery* 1: 81-126.

Beeson, K. C., V. A. Lazar, and S. G. Boyce. 1955. Some Plant Accumulators of the Micronutrient Elements. *Ecology* 36(1): 155-156.

Bradley, R., A. J. Burt, and D. J. Read. 1981. Mycorrhizal Infection and Resistance to Heavy Metal Toxicity in *Calluna vulgaris*. *Nature* 292: 335-337.

Britton, J. C. and B. Morton. 1989. *Shore Ecology of the Gulf of Mexico*. Austin: University of Texas Press.

Brooks, R. R. 1987. *Serpentine and its Vegetation: A Multidisciplinary Approach*. Portland, OR: Dioscorides Press.

Brooks, R. R., J. A. McCleave, and E. K. Schofield. 1977. Cobalt and Nickel Uptake by the Nyssaceae. *Taxon* 26(2/3): 197-201.

Brooks, R. R., R. D. Reeves, A. J. M. Baker, J. A. Rizzo, and H. Diaz Ferreira. 1988. The Brazilian Serpentine Plant Expedition (BRASPEX), 1988. *National Geographic Research* 6(2): 205-219.

Carwardine, M. 1986. *Iceland: Nature's Meeting Place*. Reykjavik, Iceland: Iceland Review.

Chambers, J. C. and R. C. Sidle. 1991. Fate of Heavy Metals in an Abandoned Lead-Zinc Tailings Pond: I. Vegetation. *J. Environ. Qual.* 20: 745-751.

Chronic, H. 1990. *Roadside Geology of Utah*. Missoula, MT: Mountain Press Publishing Company.

Craig, R. M. 1984. *Plants for Coastal Dunes*. Washington, D.C.: USDA-Soil Conservation Service. Agriculture Inf. Bulletin 460.

Duncan, W. H. and M. B. Duncan. 1987. *The Smithsonian Guide to Seaside Plants of the Gulf and Atlantic Coasts: from Louisiana to Massachusetts, Exclusive of Lower Peninsular Florida*. Washington, D.C.: Smithsonian Institution Press.

Fridriksson, S. 1975. *Surtsey: Evolution of Life on a Volcanic Island*. London: Butterworth.

Fridriksson, S. 1987. Plant Colonization of a Volcanic Island, Surtsey, Iceland. *Arctic and Alpine Research* 19(4): 425-431.

Gibson, D. J., J. S. Ely, P. B. Looney, and P. T. Gibson. 1995. Effects of Inundation from the Storm Surge of Hurricane Andrew upon Primary Succession on Dredge Spoil. *Journal of Coastal Research*, Sp. Issue #21: 208-215.

Gibson, D. J. and P. B. Looney. 1992. Seasonal Variation in Vegetation Classification on Perdido Key, a Barrier Island off the Coast of the Florida Panhandle. *Journal of Coastal Research* 8(4): 943-956.

Johnson, A. F. 1996. Vegetation Succession on Florida Dunes. Submitted for publication.

Kruckeberg, A. R. 1984. *California Serpentines: Flora, Vegetation, Geology, Soils, and Management Problems*. Berkeley: University of California Press.

Looney, P. B. and D. J. Gibson. 1993. Vegetation Monitoring of Beach Nourishment. In D. K. Stauble and N. C. Kraus, eds. *Beach Nourishment Engineering and Management Considerations*, pp. 226-241. New York: Coastlines of the World, American Society of Civil Engineers.

Mansfield, T. A., ed. 1976. *Effects of Air Pollutants on Plants*. Cambridge, Great Britain: Cambridge University Press.

Moormann, F. R. and L. J. Pons. 1974. Characteristics of Mangrove Soils in Relation to their Agricultural Land Use and Potential. In G. E. Walsh, S. C. Snedaker, and H. J. Teas, eds. *Proceedings of the International Symposium on Biology and Management of Mangroves*, vol. 2, pp. 529-547. Honolulu, Hawaii: East-West Center.

Ranwell, D. S. 1972. *Ecology of Salt Marshes and Sand Dunes*. London: Chapman and Hall Ltd.

Reeves, R. D. and R. R. Brooks. 1983. Hyperaccumulation of Lead and Zinc by Two Metallophytes from Mining Areas of Central Europe. *Environmental Pollution* (Series A) 31: 277-285.

Stokes, W. L. 1986. *Geology of Utah*. Salt Lake City, UT: Utah Museum of Natural History and Utah Geological and Mineral Survey.

Stone, C. P. and L. W. Pratt. 1994. *Hawaii's Plants and Animals: Biological Sketches of Hawaii Volcanoes National Park*. Honolulu: University of Hawaii Press.

Surrency, D. 1992. *Measures for Stabilizing Coastal Dunes*. Washington, D.C.: USDA-Soil Conservation Service.

Wilkinson, T. 1995. Land of the Moving Earth. *Backpacker* June 1995: 36-43.

5 *Salty Environments*

Arenas, V. and G. de la Lanza. 1983. Annual Phosphorus Budget of a Coastal Lagoon in the Northwest of Mexico. *Ecol. Bull.* 35: 431-440.

Austin, O. L., Jr. 1967. *Water and Marsh Birds of the World*. New York: Golden Press.

Cintrón, G. and Y. Schaeffer-Novelli. 1983. *Introducción a la Ecología del Manglar*. Montevideo, Uruguay: Oficina Regional de Ciencia y Tecnología de la Unesco para América Latina y el Caribe.

Clark, A. 1994. Samphire, From Sea to Shining Seed. *Aramco Journal* Nov.-Dec. 1994: 2-9.

Conrad, J. 1985. Mexico's Cooperative Oyster and Shrimp Farms. *Aquaculture Magazine* Sept.-Oct. 1985: 46-49.

Dodd, J. D. and R. T. Coupland. 1966. Vegetation of Saline Areas in Saskatchewan. *Ecology* 47(6): 958-967.

Foskett, J. K. and C. Scheffey. 1982. The Chloride Cell: Definitive Identification as the Salt-Secretory Cell in Teleosts. *Science* 215: 164-166.

Hammer, U. T., R. C. Haynes, J. M. Heseltine, and S. M. Swanson. 1975. The Saline Lakes of Saskatchewan. *Verh. Internat. Verein. Limnol.* 19: 589-598.

Hummelinck, M. G. W. 1984. Tidal Areas: a Blessing in Disguise. *Environment Features* 3: 1-4.

Hutchings, P. and P. Saenger. 1987. *Ecology of Mangroves*. New York: University of Queensland Press.

Jefferies, R. L. and T. Rudmik. 1984. The Responses of Halophytes to Salinity: An Ecological Perspective. In R. C. Staples and G. H. Toeniessen, eds. *Salinity Tolerance of Plants*, pp. 213-227. New York: John Wiley & Sons, Inc.

Keith, L. B. 1958. Some Effects of Increasing Soil Salinity on Plant Communities. *Canadian Journal of Botany* 36: 79-89.

LaBounty, J. F. and J. E. Deacon. 1972. *Cyprinodon milleri*, a New Species of Pupfish (Family Cyprinontidae) from Death Valley, California. *Copeia* 4: 769-780.

MacDonald, K. B. and M. G. Barbour. 1974. Alaska. In R. J. Reimold and W. H. Queen, eds. *Ecology of Halophytes*, pp. 202-206. New York: Academic Press, Inc.

McMillan, C. and F. N. Moseley. 1967. Salinity Tolerances of Five Marine Spermatophytes of Redfish Bay, Texas. *Ecology* 48(3): 503-506.

Montague, C. L., W. R. Fey, and D. M. Gillespie. 1982. A Causal Hypothesis Explaining Predator-Prey Dynamics in Great Salt Lake, Utah. *Ecological Modelling* 17: 243-270.

O'Leary, J. W. 1984. The Role of Halophytes in Irrigated Agriculture. In R. C. Staples and G. H. Toeniessen, eds. *Salinity Tolerance of Plants*, pp. 285-300. New York: John Wiley & Sons, Inc.

Pojar, J. and A. MacKinnon, eds. 1994. *Plants of Coastal British Columbia: including Washington, Oregon & Alaska*. Redmond, WA: Lone Pine Publishing.

Provasoli, L. and K. Shiraishi. 1959. Axenic Cultivation of the Brine Shrimp *Artemia salina*. *Biological Bulletin* 117: 347-355.

Schmalzer, P. A. and C. R. Hinkle. 1985. *A Brief Overview of Plant Communities and the Status of Selected Plant Species at John F. Kennedy Space Center, Florida.* A report prepared by The Bionetics Corp. for NASA.

Williams, W. D. 1985. Salt Lakes. In Twidale, C. R., M. J. Tyler, and M. Davies, eds. *Natural History of Eyre Peninsula*, pp. 119-126. Adelaide, South Australia: Royal Society of South America.

Woodroffe, C. D. 1985a. Studies of a Mangrove Basin, Tuff Crater, New Zealand: I. Mangrove Biomass and Production of Detritus. *Estuarine, Coastal and Shelf Science* 20: 265-280.

Woodroffe, C. D. 1985b. Studies of a Mangrove Basin, Tuff Crater, New Zealand: II. Comparison of Volumetric and Velocity-Area Methods of Estimating Tidal Flux. *Estuarine, Coastal and Shelf Science* 20: 431-445.

Woodroffe, C. D. 1985c. Studies of a Mangrove Basin, Tuff Crater, New Zealand: III. The Flux of Organic and Inorganic Particulate Matter. *Estuarine, Coastal and Shelf Science* 20: 447-461.

6 *Seasonal and Vernal Wetlands*

Adams, A. 1984. Cryptobiosis in Chironomidae (Diptera)—two decades on. *Antenna* 8:58-61.

Bold, H. C. 1977. 4th Ed. *The Plant Kingdom*. Englewood Cliffs, NJ: Prentice-Hall, Inc.

Cantrell, M. A. and A. J. McLachlan. 1982. Habitat Duration and Dipteran Larvae in Tropical Rain Pools. *Oikos* 38: 343-348.

Duellman, W. E. and L. Trueb. 1986. *Biology of Amphibians*. New York: McGraw-Hill Book Company.

Kinchin, J. M. 1994. *The Biology of Tardigrades*. Chapel Hill: Portland Press Inc.

Martin, G. 1990. Spring Fever. *Discover*. March: 71-74.

Matthews, E. G. 1976. *Insect Ecology*. Queensland, Australia: University of Queensland Press.

Rzòska, J. 1984. Temporary and Other Waters. In Cloudsley-Thompson, J. L., ed. *Sahara Desert*, pp. 105-114. New York: Pergamon Press.

Ward, J. V. 1992. *Aquatic Insect Ecology: 1. Biology and Habitat*. New York: John Wiley & Sons, Inc.

Wetzel, R. G. 1983. 2nd Ed. *Limnology*. Philadelphia: Saunders College Publishing.

Zedler, P. H. 1987. *The Ecology of Southern California Vernal Pools: A Community Profile*. Washington, D. C.: U.S. Fish Wildl. Serv. Biol. Rep. 85(7.11).

7 *Permanent Aquatic Environments*

Alexander, M. 1977. *Introduction to Soil Microbiology*. New York: John Wiley & Sons, Inc.

Bouldin, D. R., R. L. Johnson, C. Burda, and C. Kao. 1974. Losses of Inorganic Nitrogen from Aquatic Systems. *Journal of Environmental Quality* 3(2): 107-114.

Burckhalter, R. E. 1992. The Genus *Nyssa* (Cornaceae) in North America: A Revision. *SIDA* 15(2): 323-342.

Coles, G. C. 1970. Some Biochemical Adaptations of the Swamp Worm *Alma emini* to Low Oxygen Levels in Tropical Swamps. *Comp. Biochem. Physiol.* 34: 481-489.

Crawford, R. M. M. 1989. *Studies in Plant Survival*. Great Britain: Blackwell Scientific Publications.

Etherington, J. R. 1983. *Wetland Ecology*. London: Edward Arnold (Publishers) Ltd.

Galston, A. W. 1975. The Water Fern-Rice Connection. *Natural History* Dec. 1975: 10-11.

Gambrell, R. P. and W. H. Patrick, Jr. 1978. Chemical and Microbiological Properties of Anaerobic Soils and Sediments. In D. D. Hook and R. M. M. Crawford, eds. *Plant Life in Anaerobic Environments*, pp. 375-423. Ann Arbor, MI: Ann Arbor Science.

Harms, W. R., H. T. Schreuder, D. D. Hook, C. L. Brown, and F. W. Shropshire. 1980. The Effects of Flooding on the Swamp Forest In Lake Ocklawaha, Florida. *Ecology* 61(6): 1412-1421.

Hochachka, P. W. 1980. *Living without Oxygen: Closed and Open Systems in Hypoxia Tolerance*. Cambridge, MA: Harvard University Press.

Kawase, M. 1981. Anatomical and Morphological Adaptation of Plants to Waterlogging. *HortScience* 16(1): 30-34.

Keeley, J. E. 1979. Population Differentiation along a Flood Frequency Gradient: Physiological Adaptations to Flooding in *Nyssa sylvatica*. *Ecological Monographs* 49(1): 89-108.

Kozlowski, T. T. 1984. Plant Responses to Flooding of Soil. *BioScience* 34(3): 162-167.

McMahon, R. F. 1983. Physiological Ecology of Freshwater Pulmonates. In W. D. Russell-Hunter, ed. *The Mollusca*, pp. 404-411. Orlando, FL: Academic Press, Inc.

Munch, J. C. and J. C. G. Ottow. 1983. Reductive Transformation Mechanism of Ferric Oxides in Hydromorphic Soils. *Ecol. Bull.* 35: 383-394.

Niering, W. A. 1985. *Wetlands.* New York: Alfred A. Knopf, Inc.

Pamatmat, M. M. 1979. Anaerobic Heat Production of Bivalves (*Polymesoda caroliniana* and *Modiolus demissus*) in Relation to Temperature, Body Size, and Duration of Anoxia. *Marine Biology* 53: 223-229.

Rowell, D. L. 1981. Oxidation and Reduction. In D. J. Greenland and M. H. B. Hayes, eds. *The Chemistry of Soil Processes*, pp. 401-461. New York: John Wiley & Sons, Inc..

Saucier, J. R. 1982. *Tupelo: An American Wood.* Washington, D.C.: USDA-Forest Service. FS269.

Sommer, T. R., W. T. Potts, and N. M. Morrissy. 1990. Recent Progress in the Use of Processed Microalgae in Aquaculture. *Hydrobiologia* 204,205: 435-443.

U.S. Army Corps of Engineers Jacksonville District. 1988. *A Guide to Selected Florida Wetland Plants and Communities*. Jacksonville, FL: U.S. Army Corps of Engineers.

Vonshak, A. and A. Richmond. 1988. Mass Production of the Blue-green Algae *Spirulina*: an Overview. *Biomass* 15: 233-247.

Williams, D. M. and T. M. Embley. 1996. Microbial Diversity: Domains and Kingdoms. *Annu. Rev. Ecol. Syst.*. 27: 569-595.

Zeikus, J. G. and J. C. Ward. 1974. Methane Formation in Living Trees: A Microbial Origin. *Science* 184: 1181-1183.

8 *Polar Lands*

Allen, J. C. 1988. Averaging Functions in a Variable Environment: A Second-Order Approximation Method. *Environmental Entomology* 17(4): 623-625.

Arnett, R. H., Jr. and R. L. Jacques, Jr. 1981. *Simon & Schuster's Guide to Insects*. New York: Simon & Schuster's, Inc.

Billings, W. D. 1973. Arctic and Alpine Vegetations: Similarities, Differences, and Susceptibility to Disturbance. *Bio Science* 23(12): 697-704.

Bliss, L. C. 1971. Arctic and Alpine Plant Life Cycles. *Ann. Rev. Ecol. Syst.* 2: 405-438.

Bliss, L. C., G. M. Courtin, D. L. Pattie, R. R. Riewe, D. W. A. Whitfield, and P. Widden. 1973. Arctic Tundra Ecosystems. *Ann. Rev. Ecol. Syst.* 4: 359-399.

Bliss, L. C., G. H. R. Henry, J. Svoboda, and D. I. Bliss. 1994. Patterns of Plant Distribution within Two Polar Desert Landscapes. *Arctic and Alpine Research* 26(1): 46-55.

Bliss, L. C. and J. Svoboda. 1984. Plant Communities and Plant Production in the western Queen Elizabeth Islands. *Holarctic Ecology* 7: 325-344.

Bliss, L. C., J. Svoboda, and D. I. Bliss. 1984. Polar deserts, their Plant Cover and Plant Production in the Canadian High Arctic. *Holarctic Ecology* 7: 305-324.

Campbell, D. G. 1992. *The Crystal Desert: Summers in Antarctica.* Boston: Houghton Mifflin Company.

Edwards, J. A. 1974. Studies in *Colobanthus quitensis* (Kunth) Bartl. and *Deschampsia antarctica* Desv.: VI. Reproductive Performance on Signy Island. *Antarct. Surv. Bull.* 39: 67-86.

Edwards, J. A. 1979. An Experimental Introduction of Vascular Plants from South Georgia to the Maritime Antarctic. *Antarct. Surv. Bull.* 49: 73-80.

Edwards, J. A. and R. I. Lewis Smith. 1988. Photosynthesis and Respiration of *Colobanthus quitensis* and *Deschampsia antarctica* from the Maritime Antarctic. *Br. Antarct. Surv.* 81: 43-63.

Elliott, D. and J. Svoboda. 1994. Microecosystem around a Large Erratic Boulder: a High-Arctic Study. In J. Svoboda and B. Freedman, eds. *Ecology Of A Polar Oasis*, pp. 207-213. Ontario: Captus Press Inc.

Heinrich, B. 1990. The Antifreeze of Bees. *Natural History* July 1990: 53-58.

Klein, D. R. and C. Bay. 1991. Diet Selection by Vertebrate Herbivores in the High Arctic of Greenland. *Holarctic Ecology* 14: 152-155.

Kukal, O. 1988. Caterpillars on Ice. *Natural History* Jan. 1988: 36-40.

Kukal, O. and P. G. Kevan. 1994. The Influence of Parasitism on the Life History of a High-Arctic Insect, *Gynaephora groenlandica* (Wocke) (Lepidoptera: Lymantriidae). In J. Svoboda and B. Freedman, eds. *Ecology Of A Polar Oasis*, pp. 231-239. Ontario: Captus Press Inc.

Molau, U. 1993. Relationships between Flowering Phenology and Life History Strategies in Tundra Plants. *Arctic and Alpine Research* 25(4): 391-402.

Pielou, E. C. 1991. *After the Ice Age: The Return of Life to Glaciated North America.* Chicago: University of Chicago Press.

Pielou, E. C. 1994. *A Naturalist's Guide to the Artic.* Chicago: The University of Chicago Press.

Ross, J. 1835. Proceedings to the Tenth of April - Journey and Narrative of Commander Ross. In J. Ross. *Narrative of a Second Voyage in Search of North-West Passage*, pp. 301-674. London: A. W. Webster.

Rott, H. and F. Obleitner. 1992. The Energy Balance of Dry Tundra in West Greenland. *Arctic and Alpine Research* 24(4): 352-362.

Sage, B. 1986. *The Arctic and Its Wildlife.* New York: Facts on File Publications.

Watson, G. E. 1975. *Birds of the Antarctic and Sub-Antarctic.* Washington, D.C.: American Geophysical Union.

9 *Temperate Forests in Winter*

Kistschinski, A. A. 1976. Cold Zones. In Grzimek, B., J. Illies, and W. Klausewitz, eds. *Grzimek's Encyclopedia of Ecology*,pp. 176-186. New York: Van Nostrand Reinhold Company.

Knutson, R. M. 1979. Plants in Heat. *Natural History* 88: 42-47.

Leather, S. R., K. F. A. Walters, and J. S. Bale. 1993. Insect Cold-hardiness, ch. 4 in *The Ecology of Insect Overwintering*, pp. 75-147. Cambridge, Great Britain: Cambridge University Press.

Li, P. H. 1987. *Plant Cold Hardiness.* New York: A. R. Liss.

MacKinnon, A., J. Pojar, and R. Coupe, eds. 1992. *Plants of Northern British Columbia.* Vancouver, British Columbia: Lone Pine Publishing.

Mohlenbrock, R. H. and J. W. Voigt. 1959. *A Flora of Southern Illinois.* Carbondale: Southern Illinois University Press.

Moss, E. H. 1983. 2nd Ed., revised by J. G. Packer. *Flora of Alberta.* Toronto: University of Toronto Press.

Murray, D. F. and G. W. Douglas. 1980. The Green Mantle. In G. W. Douglas and D. F. Murray, eds. *Kluane: Pinnacle of the Yukon*, pp. 52-63. Canada: John B. Theberge.

Runkel, S. T. and A. F. Bull. 1994. *Wildflowers of Indiana Woodlands.* Ames, IA: Iowa State University Press.

Sakai, A. and W. Larcher. 1987. *Frost Survival of Plants: Responses and Adaptation to Freezing Stress.* New York: Springer-Verlag.

Stary, F. 1990. *Poisonous Plants.* Leicester, Great Britain: Magna Books.

Stefenelli, S. 1994. *Mountain Flowers of Britain and Europe.* Devon, Great Britain: David & Charles.

Storey, K. B. and J. M. Storey. 1996. Natural Freezing Survival in Animals. *Annu. Rev. Ecol. Syst.*. 27: 365-386.

Taylor, R. J. 1992. *Sagebrush Country: A Wildflower Sanctuary.* Missoula, MT: Mountain Press Publishing Company.

Vetvicka, V. 1990. *Wild Flowers of Field and Woodland.* Leicester, Great Britain: Harvey's Bookshop Ltd.

Viereck, L. A. and E. L. Little, Jr. 1991. *Alaska Trees and Shrubs.* Fairbanks: University of Alaska Press.

White, H. A., ed. 1974. *Alaska and Yukon Wild Flowers Guide.* Seattle: Alaska Northwest Books.

Wilkinson, K. 1990. *Trees and Shrubs of Alberta.* Edmonton, Alberta: Lone Pine Publishing.

10 *The Mountain Environments*

Arno, S. F. 1984. *Timberline: Mountain and Artic Forest Frontiers.* Seattle: The Mountaineers.

Azocar, A., F. Rada, and G. Goldstein. 1988. Freezing Tolerance in *Draba chionophila*, a 'miniature' Caulescent Rosette Species. *Oecologia* 75: 156-160.

Batten, D. S. and J. Svoboda. 1994. Plant Communities on the Uplands in the Vicinity of the Alexandra Fiord Lowland. In J. Svoboda and B. Freedman, eds. *Ecology Of A Polar Oasis*, pp. 97-110. Ontario: Captus Press Inc.

Beck, E. 1988. Plant Life on Top of Mt. Kilimanjaro (Tanzania). *Fiora* 181: 379-381.

Beck, E. 1994. Cold Tolerance in Tropical Alpine Plants. In Rundel, P. W., A. P. Smith, and F. C. Meinzer, eds. *Tropical Alpine Environments: Plant form and function*, pp. 77-110. Cambridge, Great Britain: Cambridge University Press.

Chronic, H. 1983. *Roadside Geology of Arizona.* Missoula, MT: Mountain Press Publishing Company.

Dahl, E. 1946. On Different Types of Unglaciated Areas during the Ice Ages and their Significance to Phytogeography. *New Phytologist* 45: 225-242.

Douglas, G. W. 1974. Montane Zone Vegetation of the Alsek River Region, southwestern Yukon. *Can. J. Bot.* 52: 2505-2532.

Fjellberg, A. 1994. Habitat Selection and Biogeography of Springtails (Collembola) from Alexandra Fiord, Ellesmere Island. In J. Svoboda and B. Freedman, eds. *Ecology Of A Polar Oasis*, pp. 227-229. Ontario: Captus Press Inc.

Fosberg, F. R. 1959. Upper Limits of Vegetation on Mauna Loa, Hawaii. *Ecology* 40(1): 144-146.

Grulke, N. E. 1995. Distribution of *Phippsia algida* and Autosuccession in the Polar Semidesert, Canadian High Arctic. *Arctic and Alpine Research* 27(2): 172-179.

Halloy, S. 1991. Islands of Life at 6000 m Altitude: The Environment of the Highest Autotrophic Communities on Earth (Socompa Volcano, Andes). *Arctic and Alpine Research* 23(3): 247-262.

Hodge, W. H. 1960. Yareta—Fuel Umbellifer of the Andean Puna. *Economic Botany* 14: 113-118.

Lauer, W. and D. Klaus. 1975. Geoecological Investigations on the Timberline of Pico De Orizaba, Mexico. *Arctic and Alpine Research* 7(4): 315-330.

Little, E. L., Jr. 1941. Alpine Flora of San Francisco Mountain, Arizona. *Madroño* 6: 65-96.

Pérez, F. L. 1987a. Needle-Ice Activity and the Distribution of Stem-Rosette Species in a Venezuelan Páramo. *Arctic and Alpine Research* 19(2): 135-153.

Pérez, F. L. 1987b. Soil Moisture and the Upper Altitudinal Limit of Giant Paramo Rosettes. *Journal of Biogeography* 14: 173-186.

Price, L. W. 1971. Vegetation, Microtopography, and Depth of Active Layer on Different Exposures in Subarctic Alpine Tundra. *Ecology* 52(4): 638-647.

Pysek, P. and J. Liska. 1991. Colonization of *Sibbaldia Tetrandra* Cushions on Alpine Scree in the Pamiro-Alai Mountains, Central Asia. *Arctic and Alpine Research* 23(3): 263-272.

Rawat, G. S. and Y. P. S. Pangtey. 1987. Floristic Structure of Snowline Vegetation in Central Himalaya, India. *Arctic and Alpine Research* 19(2): 195-201.

Rundel, P. W. 1994. Tropical Alpine Climates. In Rundel, P. W., A. P. Smith, and F. C. Meinzer, eds. *Tropical Alpine Environments: Plant form and function*, pp. 21-44. Cambridge, Great Britain: Cambridge University Press.

Schaller, G. B. and G. Binyuan. 1994. Ungulates in Northwest Tibet. *Research & Exploration* 10(3): 267-293.

Smith, A. P. and T. P. Young. 1987. Tropical Alpine Plant Ecology. *Ann. Rev. Ecol. Syst.* 18: 137-158.

Swan, L. W. 1961. The Ecology of the High Himalayas. *Scientific American* 205(4): 68-78.

Swan, L. W. 1963. Ecology of the Heights. *Natural History* 72(4): 23-29.

Walker, D. A., J. C. Halfpenny, M. D. Walker, and C. A. Wessman. 1993. Long-term Studies of Snow-Vegetation Interactions. *BioScience* 43(5): 287-301.

11 *The Environments of Mars*

Banin, A., B. C. Clark, and H. Wanke. 1992. Surface Chemistry and Mineralogy. In Kieffer, H. H., B. M. Jakosky, C. W. Snyder, and M. S. Matthews, eds. *Mars*, pp. 594-625. Tucson: University of Arizona Press.

Bold, H. C. 1978. *Introduction to the Algae*. Englewood Cliffs, NJ: Prentice-Hall, Inc.

Cloudsley-Thompson, J. L. 1975. *Terrestrial Environments*. New York: Halsted Press.

Crawford, R. M. M., H. M. Chapman, and H. Hodge. 1994. Anoxia Tolerance in High Arctic Vegetation. *Arctic and Alpine Research* 26(3): 308-312.

de Duve, C. 1995. The Beginnings of Life on Earth. *American Scientist.* 83: 428-437.

France, R. L. 1993. The Lake Hazen Trough: A Late Winter Oasis in a Polar Desert. *Biological Conservation.* 63, 149-151

Glass, B. P. 1982. *Introduction to Planetary Geology*. New York: Cambridge University Press.

Gould, S. J. 1989. *Wonderful Life: The Burgess Shale and the Nature of History*. New York: W. W. Norton & Company.

Halloy, S. and J. A. González. 1993. An Inverse Relation between Frost Survival and Atmospheric Pressure. *Arctic and Alpine Research* 25(2): 117-123.

Jakowsky, B. M. and R. M. Haberle. 1992. The Seasonal Behavior of Water on Mars. In Kieffer, H. H., B. M. Jakosky, C. W. Snyder, and M. S. Matthews, eds. *Mars*, pp. 969-1016. Tucson: University of Arizona Press.

Kennedy, A. D. 1995. Antarctic Terrestrial Ecosystem Response to Global Environmental Change. *Annu. Rev. Ecol. Syst.* 26: 683-704.

McKay, C. P., R. L. Mancinelli, C. R. Stoker, and R. A. Wharton, Jr. 1992. The Possibility of Life on Mars During a Water-Rich Past. In

Kieffer, H. H., B. M. Jakosky, C. W. Snyder, and M. S. Matthews, eds. *Mars*, pp. 1234-1245. Tucson: University of Arizona Press.

Owen, T. 1992. The Composition and Early History of the Atmosphere of Mars. In Kieffer, H. H., B. M. Jakosky, C. W. Snyder, and M. S. Matthews, eds. *Mars*, pp. 818-834. Tucson: University of Arizona Press.

Siegel, B. Z., G. McMurty, S. M. Siegel, J. Chen, and P. LaRock. 1979. Life in the Calcium Chloride Environment of Don Juan Pond, Antarctica. *Nature* 280: 828-829.

Squyres, S. W., S. M. Clifford, R. O. Kuzmin, J. R. Zimbelman, and F. M. Costard. 1992. Ice in the Martian Regolith. In Kieffer, H. H., B. M. Jakosky, C. W. Snyder, and M. S. Matthews, eds. *Mars*, pp. 523-554. Tucson: University of Arizona Press.

Stevens, T. O. and J. P. McKinley. 1995. Lithoautotrophic Microbial Ecosystems in Deep Basalt Aquifers. *Science* 270: 450-454.

Warton, R. A., G. M. Simmons, and C. P. McKay. 1989. Perenially ice-covered Lake Hoare, Antarctica: physical environment, biology and sedimentation. *Hydrobiologia* 172: 305-320.

Wayne, R. P. 1991. 2nd Ed. *Chemistry of Atmospheres*. Oxford: Clarendon Press.

Wynn-Williams, D. D. 1994. Potential Effects of Ultraviolet Radiation on Antarctic Primary Terrestrial Colonizers: Cyanobacteria, Algae, and Cryptogams. In Weiler, C. S. and P. A. Penhale, eds. *Ultraviolet Radiation in Antarctica: Measurements and Biological Effects*, pp. 243-257. Washington D. C.: American Geophysical Union.

12 *Lessons Learned*

Grove, J. 1988. *The Little Ice Age*. New York: Mathuen & Co.

Kolata, A. L. and C. R. Ortloff. 1993. Agroecological Perspectives on the Decline of the Tiwanaku State. In Kolata, A. ed. *The Tiwanaku: Portrait of an Andean Civilization*. Cambridge, MA: Blackwell.

Pearsall, D. M. 1992. The Origins of Plant Cultivation in South America. In C. Wesley and P. J. Watson, eds. *The Origins of Agriculture: An Interdisciplinary perspective*, pp. 173-205. Washington D.C.: Smithsonian Institution Press.

Stevens, G. C. and J. F. Fox. 1991. The Causes of Treeline. *Ann. Rev. Ecol. Syst.* 22: 177-191.

Wilson, L. T. and W. M. Barnett. 1983. Degree-days: An Aid in Crop and Pest Management. *California Agriculture* Jan.-Feb. 1983: 4-7.

Glossary

aa lava: a basaltic lava flow that solidifies with a jagged loose texture.
abiotic: of or characterized by the absence of life or living organisms.
aeolian: of or caused by the wind; wind-replenished.
aerenchyma: a tissue in certain aquatic plants, consisting of thin-walled cells having large intercellular spaces adapted for the circulation of air within the plant.
aestivate: to spend the summer or periods of drought in a state of torpor.
allotetraploid: an allopolyploid produced when a hybrid of two species doubles its chromosome number.
Altiplano: a plateau region in South America, situated in the Andes of Argentina, Bolivia, and Peru.
amphipod: any of numerous small, flat-bodied crustaceans of the order Amphipoda, including the beach fleas, sand hoppers, etc.
andesite: a dark-colored volcanic rock composed essentially of plagioclase feldspar and one or more mafic minerals, as hornblende or biotite.
antifreeze proteins: a class of hemolymph or serum proteins produced by several species of arthropods and polar fish; they inhibit the growth of ice crystals and should properly be termed freezing hysteresis proteins.
apomixis: any of several types of asexual reproduction, as apogamy or parthenogenesis.
appressed: pressed closely against or fitting closely to something.
aroid: any plant of the arum family.
aspartate: a salt formed from aspartic acid.

backshore: the zone of the shore or beach above the high-water line, acted upon only by severe storms or exceptionally high tides.
bajada: an alluvial plain formed at the base of a mountain by the coalescing of several alluvial fans.
barchan: a crescent-shaped sand dune with the convex side in the direction of the wind.
bentonite: a clay formed by the decomposition of volcanic ash; it has the ability to absorb large quantities of water and to expand as it does so.
biocompatibility: the capability of a substance to be in direct contact with living tissues without causing harm.
biodiversity: the existence of a wide range of different species in a given area or during a specific period of time.
biome: a complex biotic community characterized by distinctive plant and animal species and maintained under the climatic conditions of the region, esp. such a community that has developed to climax.
bromeliad: any of numerous, usually epiphytic, tropical American plants having long, stiff leaves and showy flowers; they include the pineapple,

Spanish moss, and many species grown as houseplants or ornamentals.
bulbil: a small bulb borne on the stem in place of or together with flowers, as in the viviparous knotweed.

candelillas: a couple of euphorbiaceous shrubs of the southwestern U.S. and Mexico that are the source of a wax (candelilla wax) having various commercial uses.
canthariasis: an infestation of beetles within a living mammal after larvae or adult insects enter its body.
catalase: an enzyme that decomposes hydrogen peroxide into oxygen and water.
caudex: 1. the main stem of a tree, esp. a monocot or tree fern. 2. the woody or thickened persistent base of a herbaceous perennial.
caulescent: having an obvious stem rising above the ground.
ceratopogonids: (Family Ceratopogonidae) biting midges. These midges are small to minute (in some cases less than 1 mm).
cerci: (pl. of cercus) a pair of appendages at the rear of the abdomen of certain insects and other arthropods.
Cercopithecidae: the Old World monkeys, including baboons, mandrills, and macaques; a family of mostly arboreal monkeys usually having a nonprehensile tail and four limbs almost equal in size.
chenopod: a plant belonging to the goosefoot family (Chenopodiaceae), which includes spinach and beets and many species peculiar to saline, alkaline, or desert regions.
chlorotic: of or relating to reduced levels of chlorophyll in plant leaves,causing yellowing.
coevolution: evolution involving a series of reciprocal changes in two or more noninterbreeding populations that have a close ecological relationship and act as agents of natural selection for each other, as in the succession of adaptations of a plant and its pollinator that causes them to be more interdependent on each other.
colligative: (of the properties of a substance) depending on the number of molecules or atoms rather than on their nature.
commensal: said of an animal, plant, fungus, etc., living with, on, or in another, without injury to either.
corticosteroid: any one of the steroid hormones produced by the adrenal cortex, including the glucocorticoids and the mineralocorticoids.
crassulacean: belonging to the Crassulaceae, the stonecrop family of plants.
crustose: forming a crusty, tenaciously fixed mass that covers the surface on which it grows, as certain lichens.
cryoprotective: capable of protecting against injury due to freezing, as glycerol protects frozen human cells used in medical procedures.
cultivar: a variety of plant that originated and persisted under cultivation.

cyanobacteria: blue-green algae.
cytochrome: any of several carrier molecules in the mitochondria of plant and animal cells, consisting of a protein and an iron-containing porphyrin ring and participating in the stepwise transfer of electrons in oxidation reactions: each successive cytochrome in the chain alternately accepts and releases an electron at a slightly lower energy level.
cytosol: the water-soluble components of cell cytoplasm, constituting the fluid portion that remains after removal of the organelles and other intracellular structures.

depauperate: said of an area having an impoverished fauna or flora; poorly or imperfectly developed.
diapause: a period of hormonally controlled quiescence, esp. in immature insects, characterized by cessation of growth and reduction of metabolic activity, often occurring seasonally or when environmental conditions are unfavorable.
dicot: (abrev. of dicotyledonous plant) a plant having two seed leaves.
dieoff: a sudden, natural perishing of large numbers of a species, population, or community.
dimer: a molecule composed of two identical, simpler molecules.
dimerize: to form a dimer, as in polymerization.
dinoflagellate: any of numerous chiefly marine plankton of the phylum Pyrrophyta, usually having two flagella, one in a groove around the body and the other extending from its center.
disjunct distribution: said of the geographic distribution of some plants and animals where extinction over parts of an ancestral range has left geographically isolated populations.
dytiscid beetles: the predaceous diving beetles (Family Dytiscidae).

ecophysiology: the branch of physiology that deals with the physiological processes of organisms with respect to their environment.
ecotype: a subspecies or race that is especially adapted to a particular set of environmental conditions.
ectoparasite: an external parasite (opposed to endoparasite).
edaphic: related to or caused by particular soil conditions, as of texture, chemistry or drainage, rather than by physiographic or climatic factors.
El Niño: a warm ocean current of variable intensity that develops after late December along the coast of Ecuador and Peru and sometimes causes catastrophic weather conditions.
elytra: the pair of hardened forewings of certain insects, as beetles, forming a protective covering for the flight wings; in some species the elytra are fused and the insect is effectively flightless.
embayment: a bay or a formation resembling a bay.
endodermis: in botany, a single layer of modified parenchyma cells forming a sheath around a vascular bundle.

endolithic: living embedded in the surface of rocks, as certain lichens.
endoparasite: an internal parasite (opposed to ectoparasite).
endorheic: of or pertaining to interior drainage basins.
Ephemeroptera: the mayflies, a primitive order of winged insects in the subclass Pterygota that spend most of their lives as aquatic nymphs; the fragile, membranous-winged, noneating adults emerge in large numbers and live just long enough to breed.
ephydrids: (Family Ephydridae) the shoreflies, a family of dipteran insects in the subsection Acalypteratae.
epiphyte: a plant that grows above the ground, supported nonparasitically by another plant or object, and deriving its nutrients and water from rain, the air, dust, etc.; air plant; aerophyte.
eukaryote: any organism having as its fundamental structural unit a cell type that contains specialized organelles in the cytoplasm, a membrane-bound nucleus enclosing genetic material organized into chromosomes, and an elaborate system of division by mitosis or meiosis, characteristic of all life forms except bacteria, blue-green algae, and some other microorganisms.
euphorbiaceous: belonging to the Euphorbiaceae, the spurge family of plants.
euphorb: any plant of the genus *Euphorbia*, comprising the spurges.
evaginate: to turn inside out, or cause to protrude by eversion, as a tubular organ.
evapotranspiration: the process of transferring moisture from the earth to the atmosphere by evaporation of water and transpiration from plants; the total volume transferred by this process.

flavonoid: any of a large group of aromatic oxygen heterocyclic compounds that are widely distributed in higher plants; some provide pigmentation and protection from UV radiation and others have physiologic properties.
floodplain: a nearly flat plain along the course of a stream or river that is naturally subject to flooding.
forb: any herb that is not a grass or grasslike.
fossorial: adapted for digging, as the digits, forelegs, and skeletal structure of moles, armadillos, and mole crickets.
free-living: noting an organism that is neither parasitic, symbiotic, nor sessile.
freezing-point depression: a condition in which the freezing point of a solution is lower than the standard freezing point of the solvent in a pure state; the amount of depression is directly dependent on the amount of solute present.

geobotany: the scientific study of the effects of various geological environments on plants.

geophyte: a plant propagated by means of underground buds.
glasswort: any chenopod of the genus *Salicornia*; having fleshy leafless stems and growing mainly in saltwater marshes; its soda-rich ash was formerly used in making glass.
glycolysis: the catabolism of carbohydrates, as glucose and glycogen, by enzymes, with the release of energy and the production of organic acids.
graminoid: any plant belonging to one of several families of grasslike plants.
guanidine: a colorless, crystalline, strongly alkaline, water-soluble solid, $HN{=}C(NH_2)_2$.
gypseous soils: soils containing gypsum.
gyre: a ringlike system of ocean currents rotating clockwise in the Northern Hemisphere and counterclockwise in the Southern Hemisphere.

Hadley cell: a direct, thermally driven, zonally symmetric circulation that consists of an equator-ward movement of air from 30° north or south latitude, with rising wind components near the equator, poleward flow aloft, and descending components at 30° latitude again. (It was initially proposed by George Hadley in 1735, as an explanation for trade winds.)
halobacteria: rod-shaped archaebacteria, as of the genera *Halobacterium* and *Halococcus*, occurring in saline environments as the Dead Sea, salt flats, and salt works, and using the pigment bacteriorhodopsin rather than chlorophyll for photosynthesis.
halophile: any organism, as certain halobacteria and marine bacteria, that requires a salt-rich environment for its growth and survival.
halophyte: a plant that grows in saline soils.
hemicryptophyte: a perennial plant whose buds rest on the surface of the soil hidden by scale, snow, or litter.
heterophyllly: having different kinds of leaves on the same plant.
Holarctic: belonging or pertaining to a biogeographical division comprising the Nearctic and Palearctic regions.
Humboldt Current: a cold Pacific Ocean current flowing north along the coast of Chile and Peru. Also called Peru Current.
hydathode: a specialized leaf structure through which water is exuded.
hypersaline: relating to geological material whose salinity exceeds that of seawater.
hypertonic: noting a solution of higher osmotic pressure than another solution with which it is compared; the comparison is normally made to the blood or hemolymph of an organism (opposed to hypotonic).
hypolimnion: (in certain lakes) the layer of water below the thermocline.
hypoxia: a deficiency in the amount of oxygen that reaches the tissues of an organism.
hysteresis: a lag in response exhibited by a system in reacting to changes in the forces affecting it, and varying with the system's previous history.

inquiline: an organism living in the nest, burrow, or body of another animal.
interspecies: existing or occurring between species. Also, interspecific.
intracellular: within a cell or cells.
Inuit: collectively, the Eskimo peoples inhabiting northernmost North America from northern Alaska to eastern Canada and Greenland.

limnology: the scientific study of bodies of freshwater, as lakes and ponds, with reference to their physical, geographical, biological, and other features.
lipid: any of a group of organic compounds that are greasy to the touch, insoluble in water, and soluble in alcohol and ether; lipids include fats, waxes, and other esters with analogous properties and constitute, with proteins and carbohydrates, the chief structural components of living cell membranes.
lobe-finned fish: any fish that has rounded scales and lobed fins, as the coelacanth. Also called lobefin.
lomas formations: (found in coastal deserts of South America), the plant communities found on hills or coastal ridges tall enough to receive fog deposition.

macrophyte: a plant, esp. an aquatic plant, large enough to be visible to the naked eye.
mafic: of or pertaining to rocks rich in dark, ferromagnesian minerals.
maladaptive: of, pertaining to, or characterized by maladaptation (incomplete or faulty adaptation).
malate: a salt or ester of malic acid, a dicarboxylic intermediate in the citric acid cycle.
Mediterranean climate: a local climate characterized by long, sunny, hot, dry summers and rainy winters; found on the coasts and islands of the Mediterranean Sea and also in other parts of the world, such as the Southern California coast and central Chile.
megaspores: the larger of the two kinds of spores characteristically produced by seed plants and a few fern allies, developing into a female gametophyte.
meristem: embryonic tissue in plants; undifferentiated, growing, actively dividing cells.
meromictic lake: a stratified lake that does not undergo complete mixing of its water during periods of circulation, especially a lake in which the noncirculating bottom layer does not mix with the circulating upper layer.
mesoscale: pertaining to meteorological phenomena, such as wind circulation and cloud patterns, that are about 1-100 km in horizontal extent.
metastable state: a state of a system that is in pseudo-equilibrium, such that a disturbance may disrupt the system and render it unstable;

specifically a condition in which a substance appears to be stable but has a finite probability of undergoing a spontaneous change, as when a supercooled liquid suddenly transforms into a solid. The metastable state exists at an energy level above that of a more stable state and requires the addition of a small disturbance to induce a transition to the more stable state.

microenvironment: the environment of a small area or of a particular organism; microhabitat.

microspore: the smaller of the two kinds of spores characteristically produced by seed plants and some fern allies, developing into a male gametophyte; in a flowering plant, a pollen grain.

microtopography: microrelief.

monocarpic: producing fruit only once and then dying.

montmorillonite: any of a group of clay minerals characterized by the ability to expand when they absorb large quantities of water. Also called smectite. [named after Montmorillon, France, where it was found]

mutagen: a substance or preparation capable of inducing mutation.

myiasis: the presence of fly maggots on or in the living body of a vertebrate; further classified as being facultative or obligate.

myrmecophyte: plants frequented by ants.

nm: abbreviation of nanometer (one billionth of a meter).

Neotropical: belonging or pertaining to a geographical division comprising that part of the New World extending from the tropic of Cancer southward.

nephron: the filtering and excretory unit of the kidney, consisting of the glomerulus and tubules.

nitrosamine: any of a series of compounds with the general formula R_2NNO, some of which are carcinogenic, formed in cured meats as well as other places by the reactions of amines with nitrite.

nunatak: a hill or mountain that has been completely encircled by a glacier.

nyctinasty: in higher plants, a movement associated with changes in sunlight and temperature.

oligotrophic: said of a lake characterized by a low accumulation of dissolved nutrient salts, supporting a sparse growth of algae and other organisms, and having a high oxygen content owing to the low organic content.

opuntia: a cactus of the genus *Opuntia*, characterized by leaves that are reduced to spines, flowers borne singly directly on the stem, and fruit that is a berry; some species are grown for their fruit, and to act as soil stabilizers around the world. Commonly referred to as prickly pears and chollas.

orographic effect: the rain-shadow effect produced by high mountains.

osmophile: any microorganism that thrives on or in media with high osmotic pressure.
ostracod: (seed shrimp) any of numerous tiny marine and freshwater crustaceans of the subclass Ostracoda, having a shrimplike body enclosed in a hinged bivalve shell.
oxaloacetate: a metabolic intermediate, formed from aspartic acid by transamination, that couples with acetyl-CoA to form citrate.

pahoehoe: basaltic lava having a smooth or billowy surface.
paleotropical: belonging or relating to the tropical and subtropical regions of the Eastern Hemisphere.
pantropical: living or growing throughout the tropics.
páramo: a high, cold plateau region of tropical South America.
parasitoid: an organism (usually a wasp or fly) that develops within the body of another organism (usually another insect) and eventually kills it.
perennate: to survive from season to season for an indefinite number of years.
phenology: the science dealing with the influence of climate on the recurrence of such annual phenomena of plant and animal life as budding and bird migrations.
phoresy: (among insects and arachnids) a nonparasitic relationship in which one species is carried about by another.
photic: of or pertaining to light.
photoautotroph: any organism that derives its energy for food synthesis from light and is capable of using carbon dioxide as its principal source of carbon.
photoionization: the phenomenon in which the absorption of electromagnetic radiation by an atom or molecule induces it to become ionized.
photooxidation: oxidation induced by light.
photosystem I: a series of reactions that occur during the light phase of photosynthesis, in which a pigment system absorbs photons of less than 700 nanometers and shifts the light energy to energy carriers, such as NADPH, to be used in the reduction of carbon dioxide.
photosystem II: a series of reactions that occur during the light phase of photosynthesis in which a pigment system absorbs photons of less than 685 nanometers; it is used in the photolysis or splitting of water molecules.
phreatophyte: a long-rooted plant that absorbs its water from the water table or the soil above it.
phytoplankton: the aggregate of plants and plantlike organisms in plankton.
phytotoxic: inhibitory to the growth of, or poisonous to, plants.
pneumatophore: a specialized structure developed from the root in

certain plants growing in swamps and marshes, serving as a respiratory organ.
polyhydroxy alcohol: an alcohol containing two or more hydroxyl groups and therefore able to participate in multiple hydrogen bonds.
polymerization: the combination of many small molecules to form a more complex molecule of higher molecular weight.
polyphyletic: developed from more than one ancestral type, as a group of plants or animals.
ppt: parts per thousand.
prokaryote: any cellular organism that has no nuclear membrane, no organelles in the cytoplasm except ribosomes, and has its genetic material in the form of single continuous strands forming coils or loops, characteristic of all bacteria and blue-green algae.
propagule: any structure capable of acting as an agent of reproduction for a plant or animal.
psocid: any of numerous minute winged insects of the family Psocidae (order Psocoptera), including the booklice and barklice, having mouth parts adapted for chewing and feeding on fungi, lichens, algae, decaying plant material, etc.
pterygote: belonging or pertaining to the arthropod subclass Pterygota, comprising the winged insects.
pygidium: any of various structures or regions at the caudal end of the body in certain invertebrates.

quartzitic sands: sands containing nearly 100% quartz grains.

rain forest: a tropical forest, usually of tall, densely growing, broad-leaved evergreen trees in an area of high annual rainfall.
refugium: an area with special topographic, edaphic, or climatological characteristics that remains stable within a larger area undergoing extreme, usually climatic, change, thus allowing a species or a community of species to survive after extinction in surrounding areas.
rhodopsin: photosensitive pigment, derived from vitamin A and a protein, found in the retinas of many vertebrates, including man. Its isomerization by incident light is the first step in the detection of light by the eye.

saxicolous: living or growing among rocks.
sea-blite: any herb of the genus *Suaeda*, found in salt marshes.
seagrass: any of several monocotyledonous grasslike marine plants of several genera and families; none are true grasses.
sea-milkwort: a fleshy, low, perennial herb of the primrose family, found in northern sea marshes.
selfing: a process of breeding by self-fertilization or self-pollination.
siderophore: a compound secreted by bacteria that binds ferric iron and

allows this otherwise insoluble ion to be taken up and used by the organism.
singlet oxygen: an excited state of oxygen in which all electrons are paired.
snowflush: an accumulation of drifted snow, windblown soil, and wind-transported seeds on a lee slope.
sorbitol: a naturally occurring sugar alcohol, $C_6H_8(OH)_6$; it is freely soluble in water and has a sweet taste; it is used as a food additive to enhance sweetness, and in the production of Vitamin C.
sori: (pl. of sorus) the clusters of sporangia on the back of the fronds of ferns.
speciation: the formation of new species as a result of geographic, physiological, anatomical, or behavioral factors that prevent previously interbreeding populations from breeding with each other.
sporocarp: a multicellular structure in which spores form; a fruiting body.
springtail: any of numerous minute, wingless insects of the order Collembola, most possessing a ventral forked appendage on the abdomen that is used for their characteristic springing escape.
spurry: a plant of the genus *Spergula*, of the pink family, Caryophyllaceae.
starwort: 1. any plant of the genus *Stellaria*; a chickweed. 2. the genus *Callitriche* is known as the water starworts.
stele: the central cylinder or cylinders of vascular and related tissue in the stem, root, petiole, leaf, etc. of higher plants.
sternite: a sclerite of the sternum of an insect, esp. a ventral sclerite of an abdominal segment.
stolon: a prostrate stem, at or just below the surface of the ground, that produces new plants from buds at its tips or nodes.
stoloniferous: producing or bearing stolons.
suberize: to convert into cork tissue.
subshrub: a plant consisting of a woody, perennial base with annual, herbaceous shoots; more generally, a small shrub with only partially woody stems.
supercool: to cool (a liquid) below its freezing point without producing solidification or crystallization.
superoxide: a compound containing the univalent ion O_2^-.
symbiosis: the living together of two dissimilar organisms, especially when the association is mutually beneficial.

taiga: the coniferous evergreen forests of subarctic lands, covering vast areas of northern North America and Eurasia.
taxon: a taxonomic category, as a species or genus.
tergite: the dorsal sclerite of an abdominal segment of an insect.

terricolous: living on or in the ground.
thalassohaline: said of a body of water containing salts in the same proportions as sea water.
topography: the surface features of a place or region, including hills, valleys, streams, lakes, etc.
tracheid: an elongated, tapering xylem cell having lignified, pitted, intact walls, adapted for conduction and support.
translocate: to move or transfer substances such as water or nutrients from one place to another.
trichome: an outgrowth from the epidermis of plants, as a hair.

ultramafic: (of rocks) containing iron and magnesium, with little or no silica. Also, ultrabasic.
understory: the shrubs and plants growing beneath the main canopy of a forest.
upwelling: the process by which warm, less-dense surface water is drawn away from along a shore by offshore currents and replaced by cold, denser water brought up from the subsurface.
urate: a salt of uric acid (a nearly insoluble weak acid) that is the end product of nitrogenous metabolism in many land animals.

venturi: a specially shaped constriction in a conduit that causes a pressure drop when fluid flows through it. (Named for Giovanni B. Venturi, 1746-1822, Italian physicist.)
Viking: one of a series of space probes that obtained scientific information about Mars.

weighting: the correction of measurements (usually by multiplication by predetermined factors) to account for a nonlinear relationship, as is done in calculating the heat sum or the biological aridity index.
wetland: land that has a wet and spongy soil, as a marsh, swamp, or bog.

zooplankton: the aggregate of animal or animallike organisms in plankton, as protozoans.
zooxanthellae: dinoflagellate photosynthesizing symbionts inside the cells of giant clams, flatworms, corals and other invertebrates.

Index

Order Form

If you order using this form or a reasonable facsimile, you will receive a free seed pack containing seeds mentioned in the book (while supplies last). Make sure to mention code EJS97.

◆ Fax Orders:(904) 235-2164

☎ Telephone Orders: Call Toll Free 1(888) FROGPRS [376-4777]. Have your VISA or MasterCard ready.

On-line orders: http://www.bookzone.com/bookzone/10000958.html

Postal Orders: Coquí Press, P.O. Box 32083
Panama City, FL 32407-8083, USA

❑**Please send me** ***From The Atacama to Makalu***

Name: ______________________________

Address: ______________________________

City: ____________________ State: ________ Zip: ________

Telephone: ( ) - ______________________________

Sales tax:
Please add applicable sales tax for books shipped to Florida addresses.

Shipping:
$4.00 for the first book and $2.00 for each additional book. Allow 2 weeks for delivery.

Payment:
❑Check or Money Order. Please make your check payable to Coquí Press and mail to the above address.
❑Credit card: ❑VISA ❑MasterCard

Card number: ______________________________

Authorized Signature: ____________________ Exp. date: /

Call *toll free* and order now